Teacher's guide
Mathematics

Alan Caldow and Morag McClurg

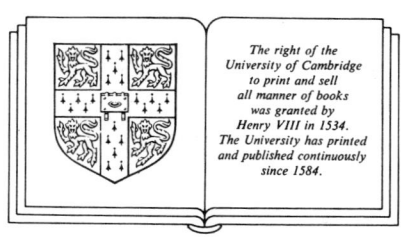

Cambridge University Press

Cambridge
New York   Port Chester   Melbourne   Sydney

Published by the Press Syndicate of the University of Cambridge
The Pitt Building, Trumpington Street, Cambridge CB2 1RP
32 East 57th Street, New York, NY 10022, USA
10 Stamford Road, Oakleigh, Melbourne 3166, Australia

© Cambridge University Press 1989

First published 1989

Printed in Great Britain at the University Press, Cambridge

British Library cataloguing in publication data
Caldow, Alan
   Teacher's guide to Mathematics for Credit 1.
   1. Mathematics
   I. McClurg, Morag   II. Caldow, Alan. Mathematics for Credit 1
   510

ISBN 0 521 36901 0

Typeset by DMD, St Clements, Oxford

Contents

	Introduction	4
1	Inequations	6
	Money management 1: pay and income tax	7
2	Square roots	7
	Money management 2: index numbers	8
3	Special angles	9
	Consolidation 1	10
4	Graphs (1)	11
	Money management 3: hire purchase and VAT	23
5	Angles greater than 90°	24
	Money management 4: savings and investment	27
6	Similar triangles	28
	Consolidation 2	29
7	The cosine rule	31
	Money management 5: loans and insurance	32
8	Patterns	33
9	Variation	34
	Consolidation 3	39

Introduction

Mathematics for Credit Books 1 and *2* have been written to provide an alternative to SMP *Books Y4* and *Y5*, thus forming a course suitable for the Credit level of the Scottish Standard Grade when used along with SMP *Books Y1, Y2* and *Y3*. In keeping with the rationale of the SMP course, an investigational approach has been taken where possible, and suitable topics are studied 'in context'.

Once the concepts of each topic are established, they are then formalised and consolidated through more traditional examples. In addition, there are sets of further consolidation exercises at suitable points in the books.

In some chapters, the early work may form the basis of class lessons or discussions; in others, a more individualised approach may be appropriate, leaving the pupils free to find their own way through the early stages.

The 'Money management' sections follow the fortunes of Cathy and Colin through the mysteries of personal finance. Cathy conveniently works for a financial consultancy that gives her advice when asked. Where possible, the examples are set in a real-life context, using true-to-life data. Often the work is investigational, requiring the pupils to make decisions about the best buy, the best loan, and so on.

As in *Books Y1* to *Y3*, there are no 'chapter summaries', so that pupils can make their own summary notes. It is felt that this is a valuable skill which should help pupils reinforce the important points of each chapter.

General notes

Calculators   It is assumed that all pupils will have access to a calculator with the basic functions, and that this will be used for all but the simplest calculations. A scientific calculator will be useful, especially for the chapters on trigonometry, but these have been written so that pupils without such a calculator will not be severely hindered. All calculations have been done to 4 s.f., with rounding to 3 s.f.

As always, pupils should be encouraged to make rough mental estimates of their answers.

Investigations Rather than include a number of free-standing investigations, we have integrated these into appropriate chapters where possible; for example, 'Graphs (1)' includes a short investigation on how the shape of the graph of a quadratic function is related to the function. Some of these investigations, especially those in which a choice has to be made, would form the basis of an interesting class discussion.

Equipment No special equipment, other than that normally found in a mathematics department, is required, nor are any worksheet masters needed.

1 Inequations

The ideas and concepts from the SMP 11–16 booklets on balancing are extended here to cope with inequations. This practical approach is then formalised and more drill is included. The idea of 'change the side, change the sign' is examined, although there is no mention of this rule in the text.

A Balances

A1 The weight of Michelle > the weight of Thomas.

A2 Balance scale is lower on the heavier side.

A3 Weight of Michelle + nappies > weight of Thomas + nappies.

A4 Weight of Savaas + cleaner < weight of Stuart + cleaner.

A5 Weight of Savaas < weight of Stuart.

B Solving simple inequations

B1 $x + 7 - 7 < 11 - 7$
$x < 4$

B2 (a) $x < 12$ (b) $x < 1$ (c) $x > 6$
(d) $x > 4$ (e) $x > -7$ (f) $x > 17$
(g) $x < 4$ (h) $x > 9$

B3 $x - 2 + 2 < 9 + 2$
$x < 11$

B4 (a) $x > 4$ (b) $x > 7$ (c) $x < 4$
(d) $x < 8$ (e) $x > -1$ (f) $x > -3$
(g) $x < -3$ (h) $x < -4$

B5 (a) $x < 5$ (b) $x < 3$ (c) $x > 5$
(d) $x > -8$ (e) $x > -6$ (f) $x < 5$
(g) $x < -4$ (h) $x > 2$ (i) $x < -6$
(j) $x < 13$ (k) $x < 16$ (l) $x > 8$

C Other 'interesting' inequations

C1 (a) 1 box > 20 kg (b) 1 barrel < 40 kg

C2 (a) $x > 7$ (b) $x < 0$ (c) $x < 10$
(d) $x > 8$ (e) $x < 16$ (f) $x > 16$
(g) $x > 10$ (h) $x < 2$ (i) $x > 54$
(j) $x < 32$ (k) $x < 2 \cdot 5$ (l) $x < 3$

D Take care with negative numbers!

D1, D2 Show your answers to your teacher.

D3 The answers are false when multiplying or dividing by a negative number is involved.

D4 (a) $x < -4$ (b) $x < 2$ (c) $x > -5$
(d) $x < -2$ (e) $x > 12$ (f) $x > -2$
(g) $x < -8$ (h) $x > 2$ (i) $x < -3$
(j) $x > 9$

D5 (a) $x > 3$ (b) $x < 2$ (c) $x > 7$
(d) $x < 10$ (e) $x < -3$ (f) $x < -5$
(g) $x < 1$ (h) $x < -4$ (i) $x < -\frac{1}{2}$
(j) $x > -\frac{1}{4}$ (k) $x < 4$ (l) $x < -4$

E Two other inequality signs

E1 (a) $x \geq 6$ (b) $x \leq 8$ (c) $x \geq 5$
(d) $x \geq 7$ (e) $x \geq 2 \cdot 6$ (f) $x \leq 9$
(g) $x \geq 2$ (h) $x \geq 4$ (i) $x \leq -4$
(j) $x \geq 2$ (k) $x \leq -2$ (l) $x \geq -2$

E2 (a) $x > 4$ (b) $x \geq 6$ (c) $x < \frac{1}{4}$
(d) $x > 0$ (e) $x \leq 2$ (f) $x \leq -12$
(g) $x < 4$ (h) $x \geq -1$ (i) $x > -1$
(j) $x \leq 10$ (k) $x > -16$ (l) $x \leq 2$

Money management 1: pay and income tax

This section begins by introducing how people's earnings can be paid in a number of different ways – per hour, per month, by piece-rate, on commission. After this, the different types of deductions from wages are examined. Income tax at the higher rates is calculated using a flow chart.

A Pay

A1 £625 **A2** £5760 **A3** £312

A4 £3·90 **A5** £279·30

A6 (a) £146·40
(b) 9 hours on Saturday, 2 hours on Sunday

B Other types of income

B1 £10 576·80

B2 (a) £192 (b) £182
(c) 40 dozen, plus 3 extra

C Deductions

C1 (a) £6864 (b) £14·75

C2 (a) £11 038 (b) £663·01

C3 (a) Sandy
(b) Flora pays £3510, Rose pays £636·07, Fred pays £24.30.
(c) Yes
(d) Show your answer to your teacher.

D Higher tax rates

D1 (a) £372 701·20 (b) £9108

D2 (a) £16 908 (b) £14 159·25
(c) £7479·15 (d) £1 525 740·60

D3 Sue, by £749·33

2 Square roots

This chapter starts by reinforcing the idea that square roots should not be approximated when they will be used in a subsequent calculation. Equivalent surds are then introduced by looking at enlargements of a given right-angled triangle, from which the pupils deduce that $\sqrt{x}\sqrt{y} = \sqrt{xy}$ and work out how to rationalise the denominator of a fractional surd. Towards the end of the chapter, the iterative method of evaluating square roots is introduced and, as a slight diversion, the rote method is also examined, but this is for interest only.

A Calculations with square roots

A1 (a) $FH = \sqrt{50}$ cm
(b) $AF = \sqrt{75} \approx 8.66$ cm

A2 9·49 cm ($\sqrt{90}$)

B Surds

B1 (a) $\sqrt{3}, \sqrt{6}, \sqrt{90}, \sqrt{203}, \sqrt{2}$
(b) $\sqrt{36}, \sqrt{49}, \sqrt{900}, \sqrt{169}, \sqrt{9}$

C Equivalent surds: products and factors

C1 (a) AC = $\sqrt{10}$ m
   (b) (i) $\sqrt{20}$ m (ii) $\sqrt{2}\sqrt{10}$
   (c) (i) $\sqrt{40}$ m (ii) $\sqrt{40} = 2\sqrt{10}$

C2 (a) $\sqrt{12}$ (b) $\sqrt{21}$ (c) $\sqrt{18}$ (d) $\sqrt{30}$
   (e) $\sqrt{30}$ (f) $\sqrt{85}$ (g) $\sqrt{180}$
   (h) 6 (i) 10 (j) 30 (k) $\sqrt{481}$
   (l) $\sqrt{2160}$

C3 (b) $\sqrt{72} = 6\sqrt{2}$ (c) $10\sqrt{5}$ (d) $6\sqrt{10}$
   (e) $7\sqrt{5}$ (f) $16\sqrt{2}$ (g) $12\sqrt{3}$
   (h) $8\sqrt{6}$ (i) $18\sqrt{3}$

C4 (a) $\sqrt{7} + 3\sqrt{3}$ (b) $8\sqrt{5} + \sqrt{3}$
   (c) $\sqrt{11} - 3\sqrt{3}$ (d) $-3\sqrt{5} - 3\sqrt{2}$
   (e) $44\sqrt{2} + 2\sqrt{5}$
   (f) $14\sqrt{2} - 6\sqrt{6} + 4\sqrt{3}$

D Equivalent surds: fractions

D1 (a) $2\sqrt{2}$ m (b) MC = $\frac{\sqrt{8}}{2}$ or $\sqrt{2}$ m

D2 $\frac{2}{\sqrt{2}}$ m

D3 MC = $\sqrt{2}$ or $\frac{\sqrt{8}}{2} = \frac{2}{\sqrt{2}}$ m

D4 $(\sqrt{2})^2 = 2$
   $\left(\frac{2}{\sqrt{2}}\right)^2 = \frac{4}{2} = 2$

D5 (a) All equal to $\frac{3}{4}$ (b) All equal to $\frac{1}{5}$
   (c) All equal to $\frac{2}{3}$

D6 (a) $\frac{\sqrt{3}}{3}$ (b) $\frac{\sqrt{6}}{3}$ (c) $\sqrt{7}$ (d) $\frac{\sqrt{6}}{3}$
   (e) $\frac{\sqrt{15}}{15}$ (f) $\frac{\sqrt{2}}{10}$ (g) $\sqrt{3}$ (h) $\frac{8\sqrt{17}}{17}$
   (i) $\frac{\sqrt{10}}{10}$ (j) $\frac{16\sqrt{5}}{5}$

E Miscellany

E1 $5\sqrt{7}$ **E2** $\sqrt{14}$ **E3** $3\sqrt{2}$
E4 $4\sqrt{10}$ **E5** $\sqrt{3}$ **E6** $5\sqrt{3} - 2\sqrt{5}$
E7 48 **E8** $\frac{1}{3}$ **E9** 1
E10 $\frac{7\sqrt{10}}{10}$ **E11** $\frac{4\sqrt{2} + 3}{2}$
E12 $\frac{17\sqrt{3} - 5}{3}$ **E13** $\frac{6 - \sqrt{7}}{7}$
E14 $\frac{4 + \sqrt{6}}{3}$ **E15** $2\sqrt{10} + 6$
E16 $12 + 2\sqrt{6}$

F Evaluating square roots

F1 (a) 3·606 (b) 2·828 (c) 5·385
   (d) 8·888 (e) 11·832 (f) 14·765

F2 (a) 42 (b) 285 (c) 586 (d) 743

F3 Show your answer to your teacher.

Money management 2: index numbers

Here we look at how proportion can be used in the context of calculating and using index numbers. Using index numbers for comparison is also examined. A further extension of this topic would be to look at the working of indexes which are in general use, such as the *Financial Times* Shares Index or the Retail Prices Index.

A Calculating index numbers

A1

Year	1983	1984	1985	1986	1987
Index	90	90	100	120	130

A2

Year	1984	1985	1986	1987	1988
Index	71	77	86	91	100

A3

Year	1985	1986	1987	1988	1989
Index	88	94	100	97	110

B Calculating prices

B1

Year	1985	1986	1987	1988
Price	49p	50p	52p	56p

B2

Year	1998	1999	2000	2001	2002
Price	£1·50	£1·65	£1·74	£1·86	£2·04

B3

Year	1987	1988	1989	1990
Price	£3·19	£3·45	£3·64	£3·75

C Using index numbers

C1 Skirt and training shoes

C2 (a) £14 040   (b) Between 1900 and 1991   (c) 1992

C3 (a)

Year	1975	1980	1985	1990	1995	2000
UK	94	96	98	100	102	105
USA	87	91	96	100	104	107

C3 (b)

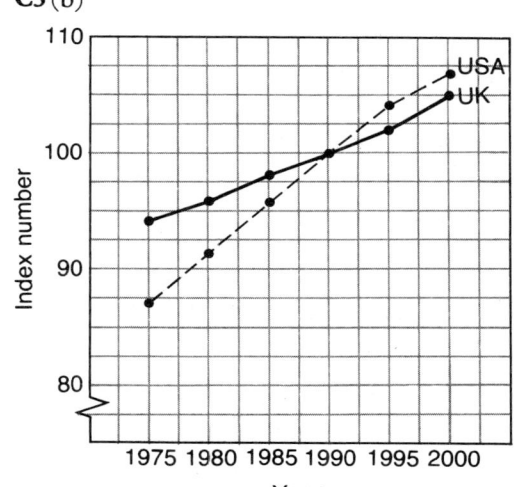

(c) Population growth in both countries is steady until 1995, when the population of the UK rises more quickly and that of the USA rises more slowly. Up until then, the population of the USA is growing more rapidly than that of the UK.

3 Special angles

This chapter continues the topic of surds, here used within a context. Pupils work out the lengths of the sides of the 'set-square' triangles and use these to find the trigonometric ratios of 30°, 45° and 60°. Identities are also introduced in an investigative way.

A Exact values of trigonometric ratios

A1 (a) BD = $\sqrt{2}$ cm   (b)

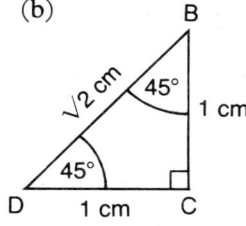

(c) $\sin 45° = \dfrac{1}{\sqrt{2}}$, $\cos 45° = \dfrac{1}{\sqrt{2}}$, $\tan 45° = 1$

A2 (a) AD = $\sqrt{3}$ cm   (b)

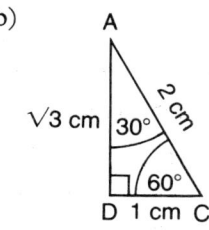

(c) $\sin 30° = \dfrac{1}{2}$, $\sin 60° = \dfrac{\sqrt{3}}{2}$

$\cos 30° = \dfrac{\sqrt{3}}{2}$, $\cos 60° = \dfrac{1}{2}$

$\tan 30° = \dfrac{1}{\sqrt{3}}$, $\tan 60° = \sqrt{3}$

A3

	30°	45°	60°
sine	$\frac{1}{2}$	$\frac{1}{\sqrt{2}}$	$\frac{\sqrt{3}}{2}$
cosine	$\frac{\sqrt{3}}{2}$	$\frac{1}{\sqrt{2}}$	$\frac{1}{2}$
tangent	$\frac{1}{\sqrt{3}}$	1	$\sqrt{3}$

B Properties of angles: an investigation

B1 (a) $\frac{1}{2}$ (b) $\frac{1}{4}$ (c) $\frac{\sqrt{3}}{2}$ (d) $\frac{3}{4}$ (e) 1

B2 45°: (a) $\frac{1}{\sqrt{2}}$ (b) $\frac{1}{2}$ (c) $\frac{1}{\sqrt{2}}$ (d) $\frac{1}{2}$
 (e) 1
 60°: (a) $\frac{\sqrt{3}}{2}$ (b) $\frac{3}{4}$ (c) $\frac{1}{2}$ (d) $\frac{1}{4}$
 (e) 1

B3 $\sin^2 x° + \cos^2 x° = 1$

B4 (a) $\frac{1}{2}, \frac{\sqrt{3}}{2}, \frac{1}{\sqrt{3}}$

 (b) $\frac{\sin 30°}{\cos 30°} = \tan 30°$

B5, B6 Show your answers to your teacher.

C Using exact values

C1 20 m

C2 (a) $\frac{14}{\sqrt{3}}$ or $\frac{14\sqrt{3}}{3}$ cm (b) 5·9 m (c) 60°

C3 $\frac{120 + 95\sqrt{3}}{2}$ m (b) $\frac{120\sqrt{3} + 95}{2}$ m

C4 (a)

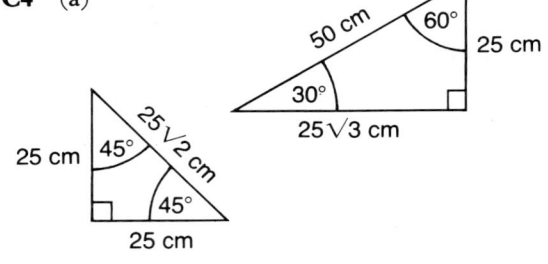

(b) $\frac{625\sqrt{3} - 625}{2}$ cm²

C5 (a) (b) 1000 m

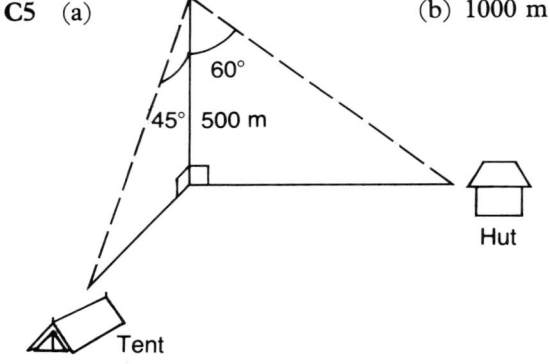

Consolidation 1

Inequations

1 (a) $x > 8$ (b) $x \leq 26$ (c) $x \geq -13$
 (d) $x < 10$ (e) $x \leq -2$ (f) $x > -2$
 (g) $x \geq 2$ (h) $x > 40$ (i) $x \leq 3$
 (j) $x < -6$ (k) $x \leq -3$ (l) $x < -4$
 (m) $x \geq -0·5$ (n) $x > -3$ (o) $x \geq -2$
 (p) $x \geq 4$ (q) $x \geq \frac{1}{4}$ (r) $x < -\frac{1}{2}$
 (s) $x > 12$ (t) $x \geq -8$

2 $x + 1·9 < 7·2$
  $x < 5·3$

Pay and income tax

1 £8220 **2** £460 **3** £91·24

4 (a) £35 700 (b) £11 074·32
 (c) £20 787·93

5 (a) £228 (b) £185·25
 (c) Minimum 540, maximum 551
 (d) £153·67, £134·86, £163·07

Square roots

1 $2\sqrt{82}$ cm

2 (a) $2\sqrt{3}$ (b) $10\sqrt{5}$ (c) $23\sqrt{3}$
   (d) $17\sqrt{2}$ (e) 9 (f) $4\sqrt{23}$

3 (a) $\dfrac{5\sqrt{6}}{6}$ (b) $\sqrt{6}$ (c) $\dfrac{7\sqrt{5}}{10}$ (d) $\dfrac{\sqrt{3}}{3}$
   (e) $\dfrac{130}{104}$

4 (a) 3·464 (b) 5·745 (c) 4·899

Index numbers

1 (a)

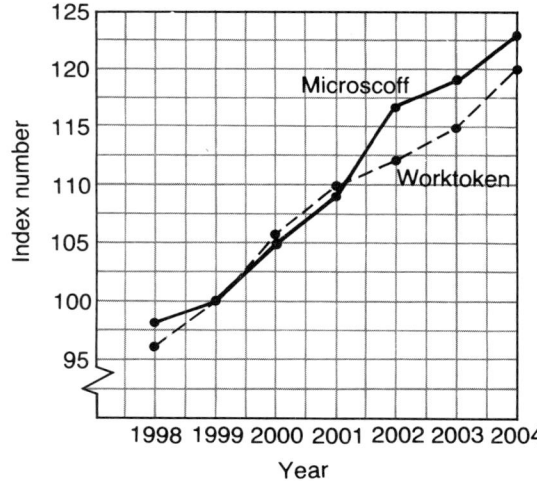

1 (b) 2001 (c) 17% (d) 300 megapounds
   (e) Between 2001 and 2002

2
Year	1998	1999	2000	2001
Index	94	100	109	123

Special angles

1 5 cm   2 $\dfrac{20\sqrt{3}}{3}$ cm, $\dfrac{100\sqrt{3}}{3}$ cm^2

3 cos 60°, cos 45°, sin 60°, tan 45°

4 $25\sqrt{2}$ m   5 $4\sqrt{3}$ km

Graphs (1)

The idea of being able to explain a rule both in words and in symbols is critical to the section on functional notation. The chapter builds on the foundation laid in earlier work on number machines and leads into more formal work on functional notation. After practice in changing from one form to another, the rest of the chapter concentrates on drawing graphs by making up a table of values and on using graphs to solve real-life problems.

Material from the Shell package could be used to supplement the chapter with more investigations.

A Number machines

A1 (a) 28 (b) 18 (c) −12

A2 (a) 3 (b) 143 (c) 0

A3 (a) Because $\sqrt{9} = 3$ or -3
(b) (i) 8 or −2 (ii) 3 (iii) 10 or −4

A4 (a) (i) 5 or −5 (ii) 2·5 or −2·5
  (iii) 6 or −6
(b) No answer is possible, since we cannot find the square root of a negative number.

A5 (a), (b), (d) and (e) are functions.
(c) is not a function because negative numbers cannot go in.
(f) is not a function because more than one answer at a time comes out.

A6 (a) $7x - 3$ (b) $a^2 + 1$ (c) $6(m + 4)$
(d) $\dfrac{p - 3}{5}$ (e) $\dfrac{q}{5} - 3$

B Functional notation

B1 (a) $f(x) = 4x - 1$ (b) $g(a) = 5a + 7$
(c) $h(b) = \dfrac{b}{3} + 1$ (d) $m(y) = 3(y - 2)$
(e) $n(z) = \dfrac{(z + 4)}{6}$

B2 (a) $f(x) = 2x - 7$ (b) $f(x) = x^2 + 7$
(c) $f(x) = 2(x + 1)$ (d) $f(x) = 3(x^2 - 7)$

B3 (a) Add 7
(b) Multiply by itself, then subtract from 4
(c) Multiply by itself, subtract 6, then divide by 2
(d) Multiply by itself, multiply by 3, then subtract 1
(e) Multiply by itself, add 2, then multiply by 9

C Evaluations

C1 (a) 15 (b) −5 (c) −17

C2 (a) 5 (b) −9 (c) 11

C3 (a) 2 (b) 18 (c) 3

C4 (a) 3 (b) 9·8 (c) −9 (d) 56

D Linear functions

D1 (a)

x	1	2	3	4	5	6
$f(x)$	3	7	11	15	19	23

(b)

x	0	1	2	3	4	5	6
$g(x)$	3	5	7	9	11	13	15

(c)

x	0	1	2	3	4	5
$h(x)$	3	−3	−9	−15	−21	−27

(d)

x	1	2	3	4	5
$p(x)$	6	10	14	18	22

(e)

x	0	1	2	3	4	5
$v(x)$	−1	$\tfrac{1}{2}$	2	$\tfrac{7}{2}$	5	$\tfrac{13}{2}$

(f)

x	0	1	2	3	4	5
$q(x)$	$\tfrac{1}{2}$	$-3\tfrac{1}{2}$	$-7\tfrac{1}{2}$	$-11\tfrac{1}{2}$	$-15\tfrac{1}{2}$	$-19\tfrac{1}{2}$

D2

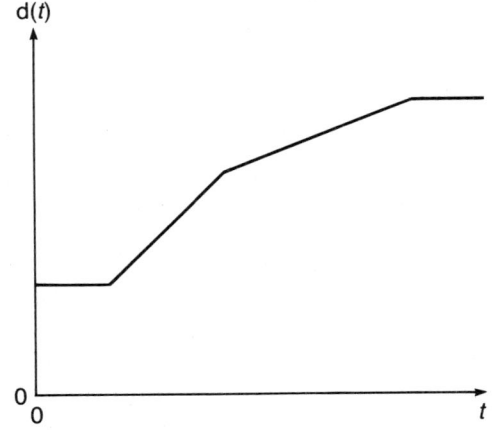

D3

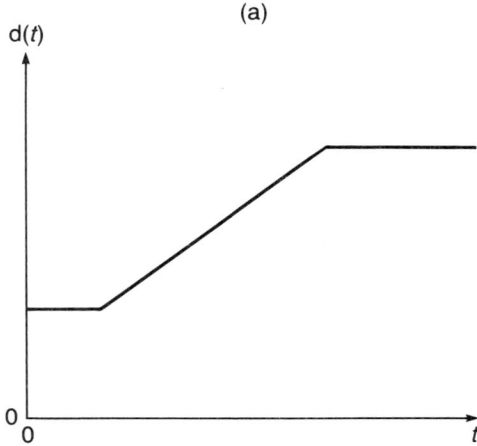

(a)

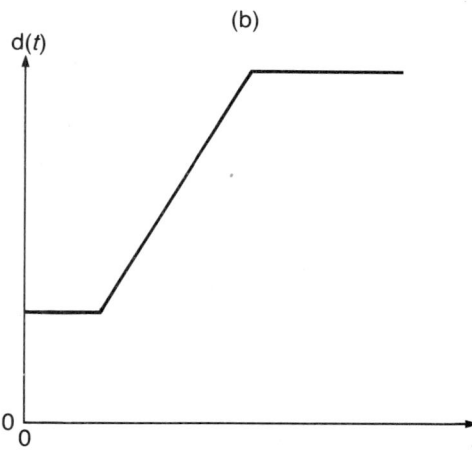

(b)

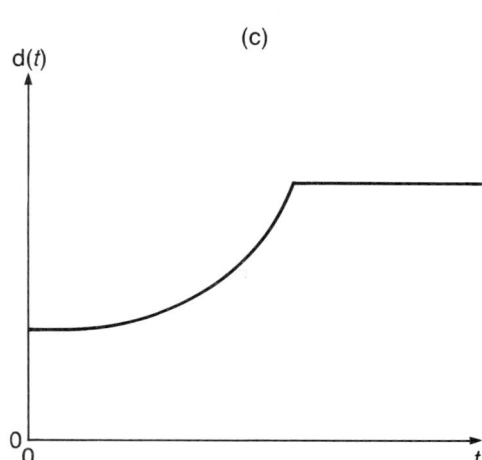

(c)

D4 Show your answer to your teacher.

D5 (a)

x	0	1	2	3	4	5
$f(x)$	3	7	11	15	19	23

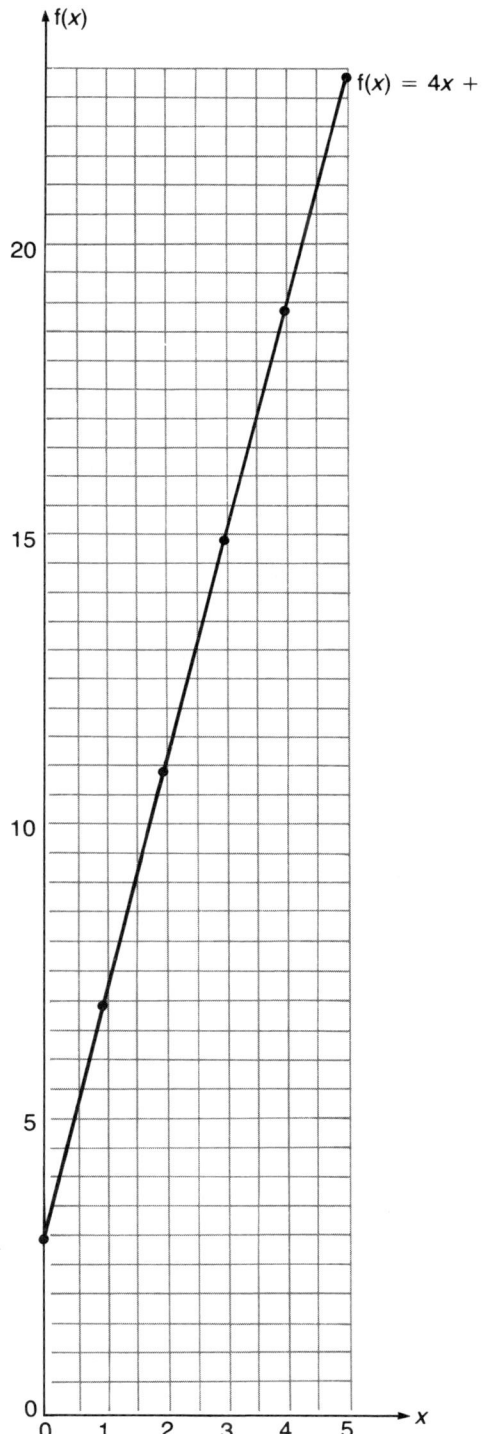

D5 (b)

x	0	1	2	3	4	5
$g(x)$	−1	2	5	8	11	14

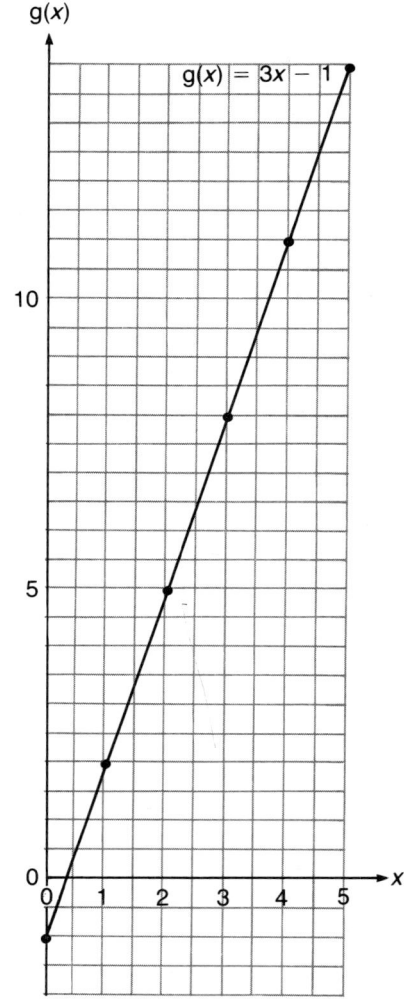

D5 (c)

x	0	1	2	3	4	5
$c(x)$	5	4	3	2	1	0

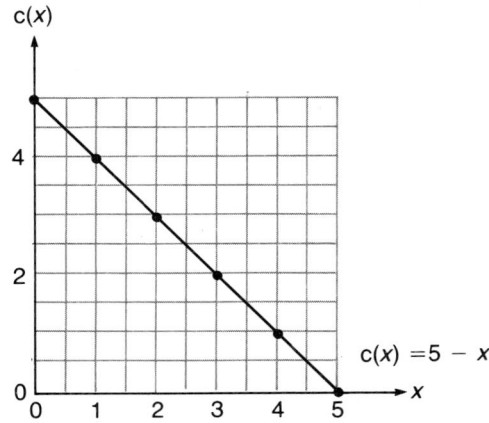

D5 (d)

x	−4	−3	−2	−1	0	1	2
$h(x)$	9	7	5	3	1	−1	−3

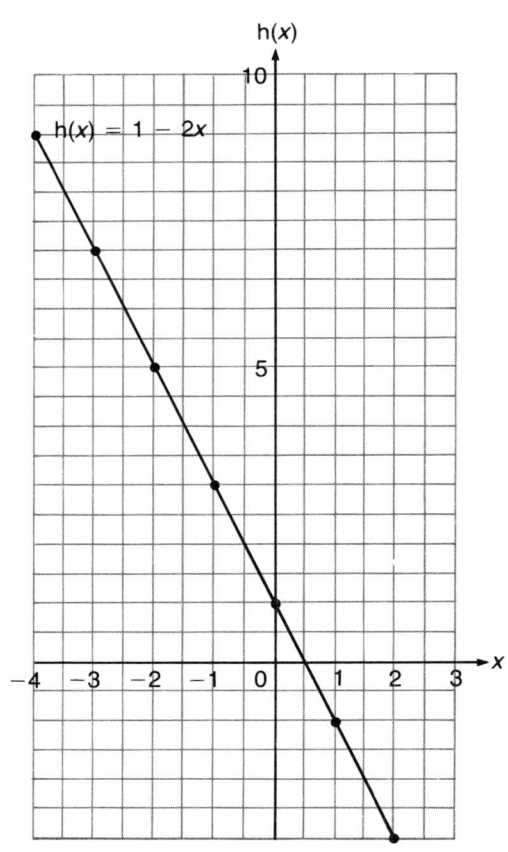

D5 (e)

x	−3	−2	−1	0	1	2
p(x)	−20	−16	−12	−8	−4	0

D5 (f)

x	−2	−1	0	1	2	3	4
q(x)	1	$\frac{3}{2}$	2	$\frac{5}{2}$	3	$\frac{7}{2}$	4

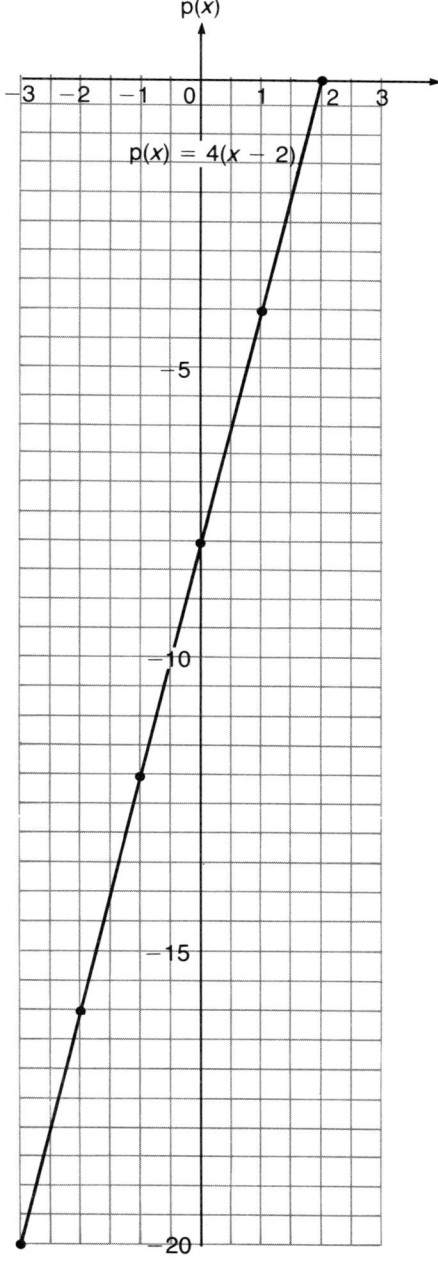

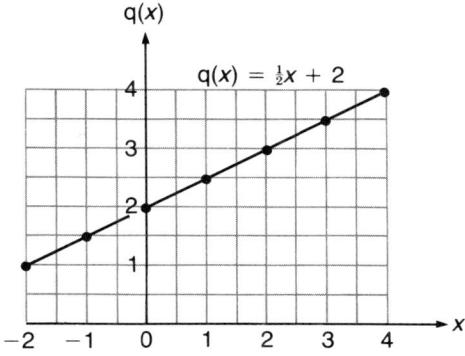

E Quadratic functions

E1 (a), (d) and (e) are quadratic functions.

F Evaluating quadratic functions

F1 (a)

x	−2	−1	0	1	2	3	4	5	6
x^2	4	1	0	1	4	9	16	25	36
$-2x$	4	2	0	−2	−4	−6	−8	−10	−12
-8	−8	−8	−8	−8	−8	−8	−8	−8	−8
f(x)	0	−5	−8	−9	−8	−5	0	7	16

(b) (−2, 0), (−1, −5), (0, −8), (1, −9), (2, −8), (3, −5), (4, 0), (5, 7), (6, 16)

(c)

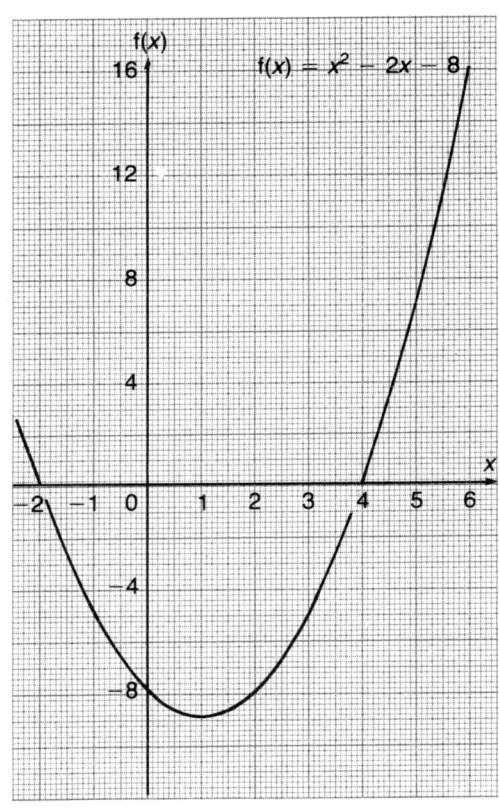

F2 (a)

x	−3	−2	−1	0	1	2	3	4
x^2	9	4	1	0	1	4	9	16
$-2x$	6	4	2	0	−2	−4	−6	−8
-3	−3	−3	−3	−3	−3	−3	−3	−3
f(x)	12	5	0	−3	−4	−3	0	5

(b) (−3, 12), (−2, 5), (−1, 0), (0, −3), (1, −4), (2, −3), (3, 0), (4, 5)

(c)

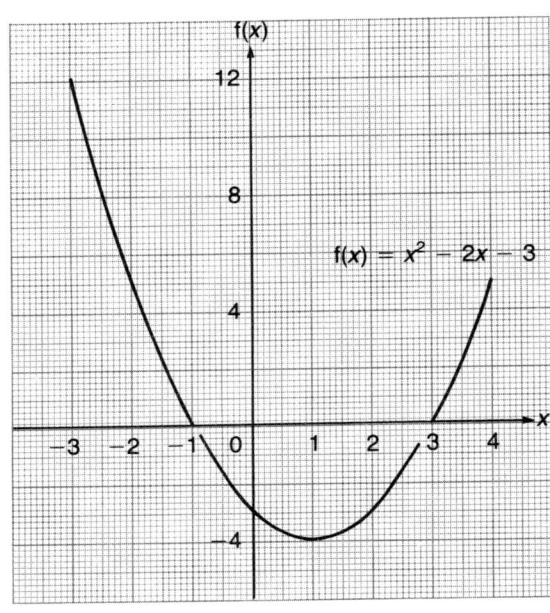

F3 (a)

x	−4	−3	−2	−1	0	1	2	3	4
9	9	9	9	9	9	9	9	9	9
$-x^2$	−16	−9	−4	−1	0	−1	−4	−9	−16
$f(x)$	−7	0	5	8	9	8	5	0	−7

(b) (−4, −7), (−3, 0), (−2, 5), (−1, 8), (0, 9), (1, 8), (2, 5), (3, 0), (4, −7)

(c)

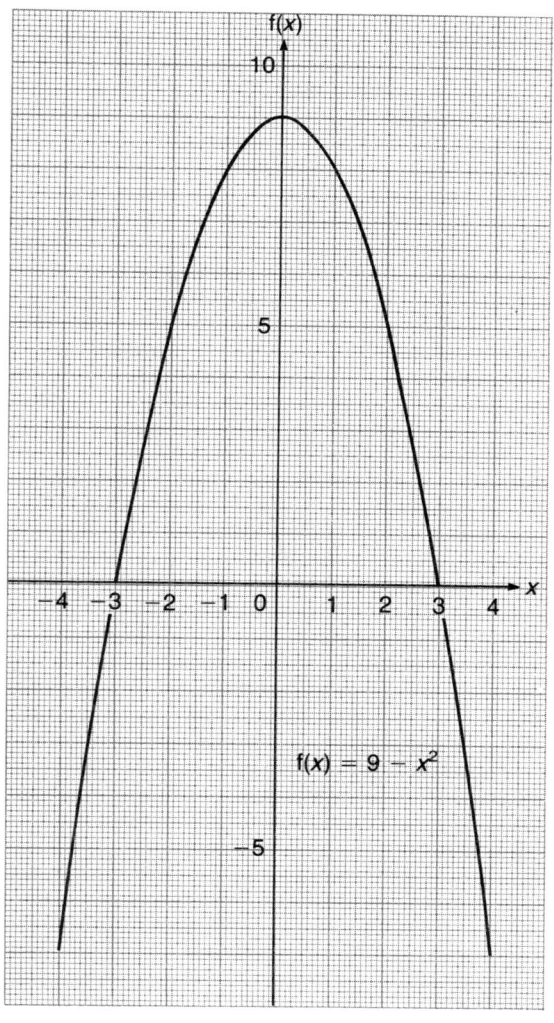

F4 (a)

x	−3	−2	−1	0	1	2
$-x^2$	−9	−4	−1	0	−1	−4
$-x$	3	2	1	0	−1	−2
2	2	2	2	2	2	2
$f(x)$	−4	0	2	2	0	−4

(b) (−3, −4), (−2, 0), (−1, 2), (0, 2), (1, 0), (2, −4)

(c)

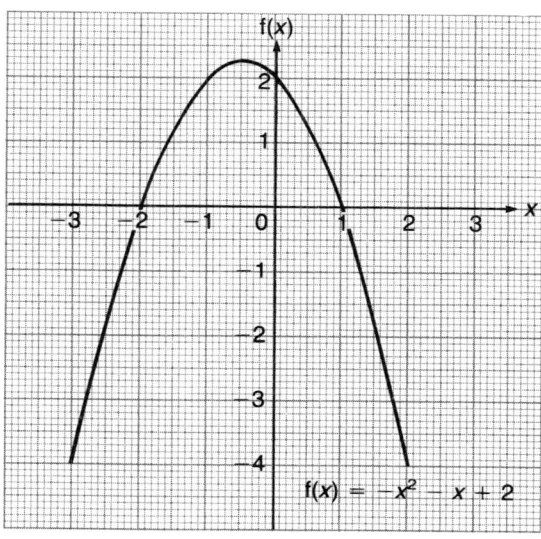

F5 (a)

x	-1	0	1	2	3
$-2x^2$	-2	0	-2	-8	-18
$7x$	-7	0	7	14	21
-6	-6	-6	-6	-6	-6
$f(x)$	-15	-6	-1	0	-3

(b) $(-1, -15)$, $(0, -6)$, $(1, -1)$, $(2, 0)$, $(3, -3)$

(c)

F6 (a)

x	-2	-1	0	1	2
$2x^2$	8	2	0	2	8
$3x$	-6	-3	0	3	6
3	3	3	3	3	3
$f(x)$	5	2	3	8	17

(b) $(-2, 5)$, $(-1, 2)$, $(0, 3)$, $(1, 8)$, $(2, 17)$

(c)

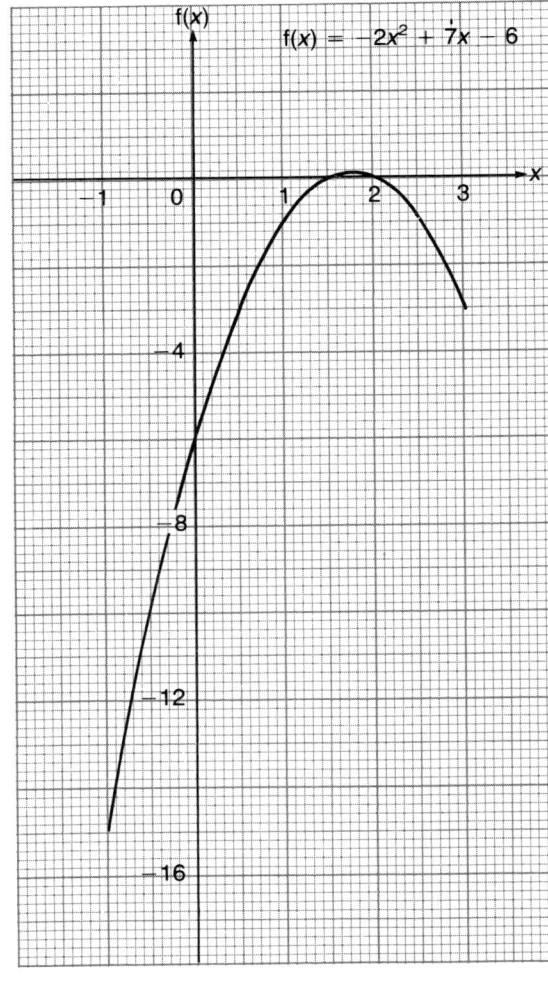

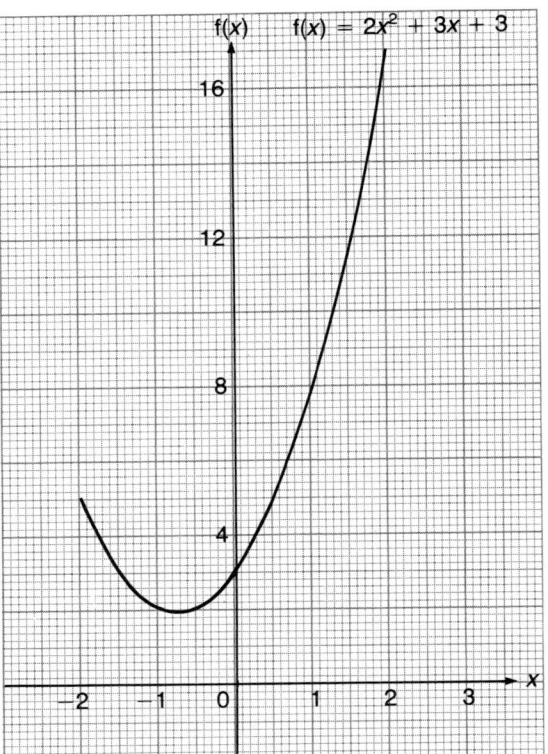

F7

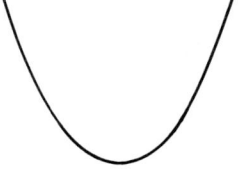

F8

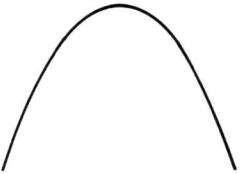

18

F9 $f(x) = x^2 - 2x - 3$: roots at $x = -1$ and $x = 3$
$f(x) = 9 - x^2$: roots at $x = -3$ and $x = 3$
$f(x) = -x^2 - x + 2$: roots at $x = -2$ and $x = 1$
$f(x) = -2x^2 + 7x - 6$: roots at $x = 2$ and $x = \frac{3}{2}$
$f(x) = 2x^2 + 3x + 3$: no roots

F10 Where it cuts the x-axis

F11 (a) Lowest (b) Minimum (c) −9
(d) $x = 1$
(e) Half-way between the roots

F12

Function	Maximum or minimum turning point?	Turning value	Coordinates of turning point
$f(x) = x^2 - 2x - 3$	Minimum	−4	$(1, -4)$
$f(x) = 9 - x^2$	Maximum	9	$(0, 9)$
$f(x) = -x^2 - x + 2$	Maximum	$2\frac{1}{4}$	$(-\frac{1}{2}, 2\frac{1}{4})$
$f(x) = -2x^2 + 7x - 6$	Maximum	$\frac{1}{8}$	$(\frac{7}{4}, \frac{1}{8})$
$f(x) = 2x^2 + 3x + 3$	Minimum	$1\frac{7}{8}$	$(-\frac{3}{4}, 1\frac{7}{8})$

G Investigation

Show your answers to your teacher.

H Making the most of it

H1 (a) $A(x) = 20x - x^2$

(b)
x	0	4	8	12	16	20
20x	0	80	160	240	320	400
−x²	0	−16	−64	−144	−256	−400
A(x)	0	64	96	96	64	0

(c)

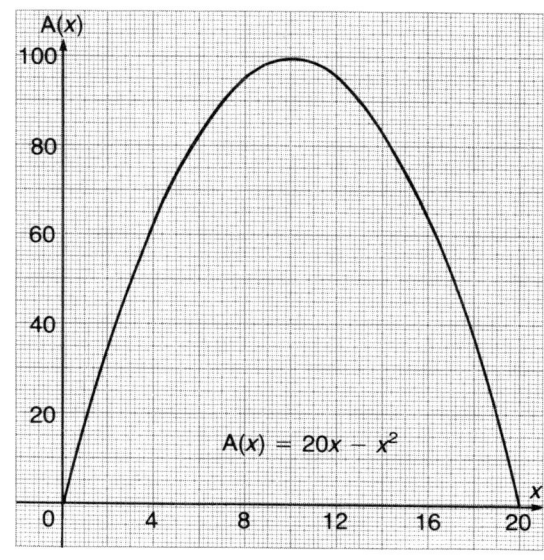

(d) $x = 10$, area $= 100$ m²

H2 (a) $A(x) = 50x - 2x^2$

(b)

x	0	5	10	15	20	25
$50x$	0	250	500	750	1000	1250
$-2x^2$	0	-50	-200	-450	-800	-1250
$A(x)$	0	200	300	300	200	0

(c)

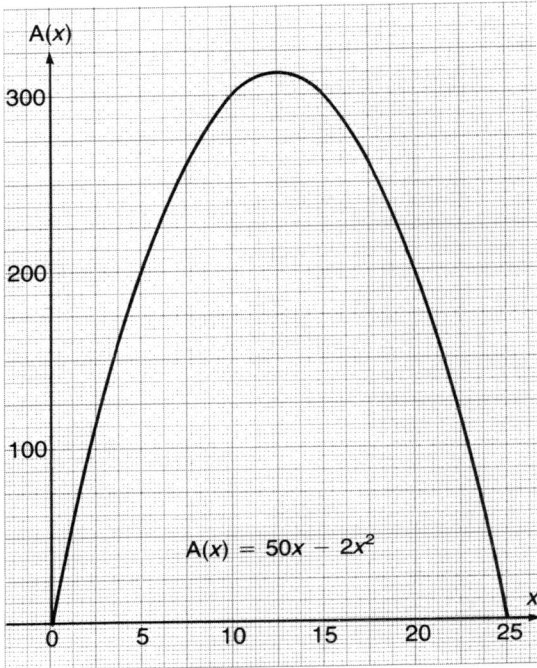

(d) $312{\cdot}5$ m^2

H3 Against the wall
Length 5 m, breadth $2{\cdot}5$ m, area $12{\cdot}5$ m^2

H4 No, it falls 30 m short.

H5 (a) 5625 m (b) 30 min

I Cubic functions

I1 (a)

x	-1	0	1	2	3	4	5	6
x^3	-1	0	1	8	27	64	125	216
$-8x^2$	-8	0	-8	-32	-72	-128	-200	-288
$15x$	-15	0	15	30	45	60	75	90
$f(x)$	-24	0	8	6	0	-4	0	18

(b)

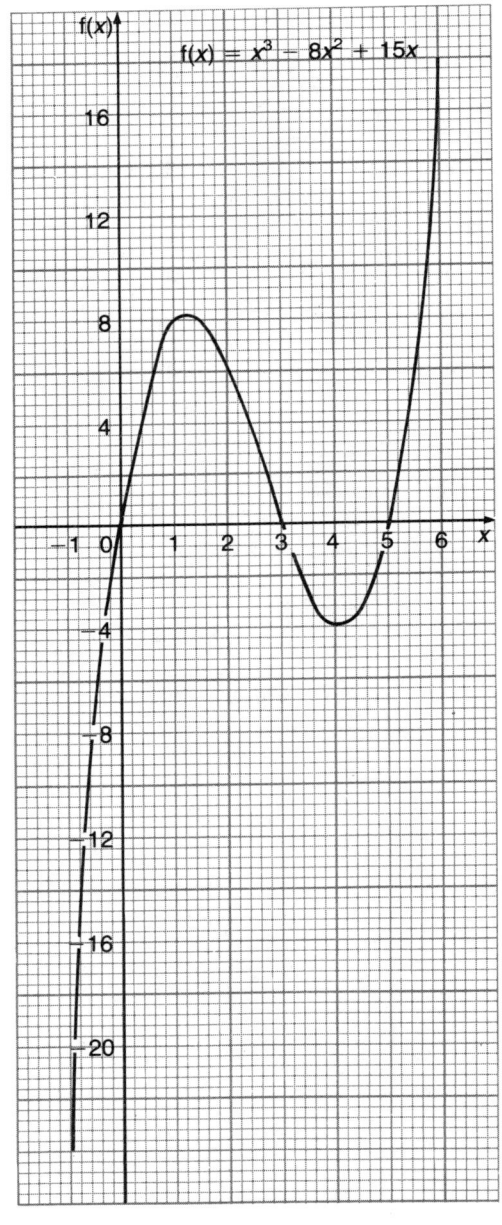

(c) Roots at $x = 0$, $x = 3$ and $x = 5$

I2 (a)

x	-3	-2	-1	0	1	2	3	4
x^3	-27	-8	-1	0	1	8	27	64
$-2x^2$	-18	-8	-2	0	-2	-8	-18	-32
$-5x$	15	10	5	0	-5	-10	-15	-20
$+6$	6	6	6	6	6	6	6	6
$f(x)$	-24	0	8	6	0	-4	0	18

(b)

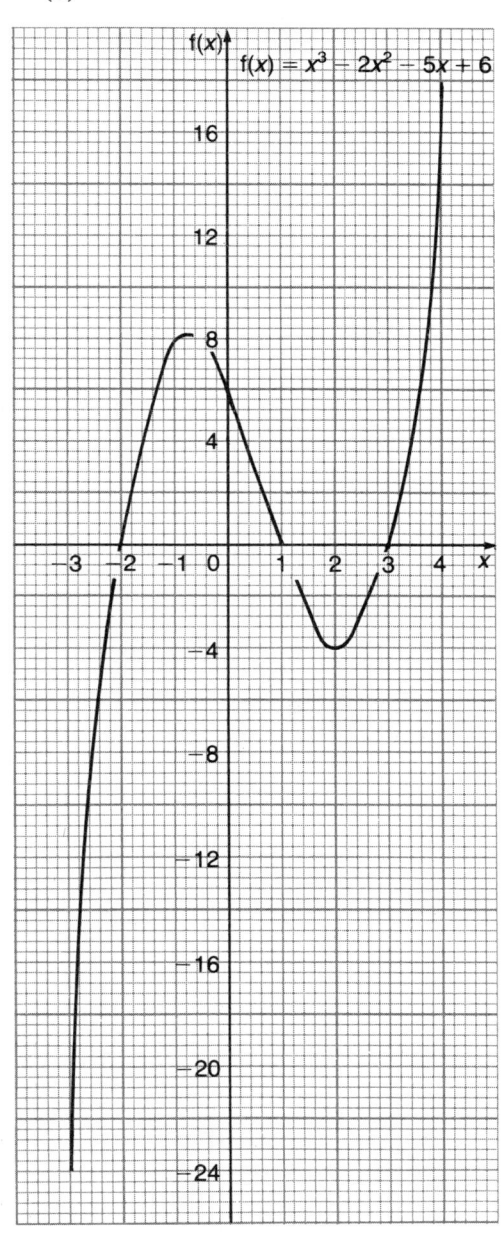

(c) Roots at $x = -2$, $x = 1$ and $x = 3$

I3 (a)

x	-3	-2	-1	0	1	2
$2x^3$	-54	-16	-2	0	2	16
$3x^2$	27	12	3	0	3	12
$-3x$	9	6	3	0	-3	-6
-2	-2	-2	-2	-2	-2	-2
$f(x)$	-20	0	2	-2	0	20

(b)

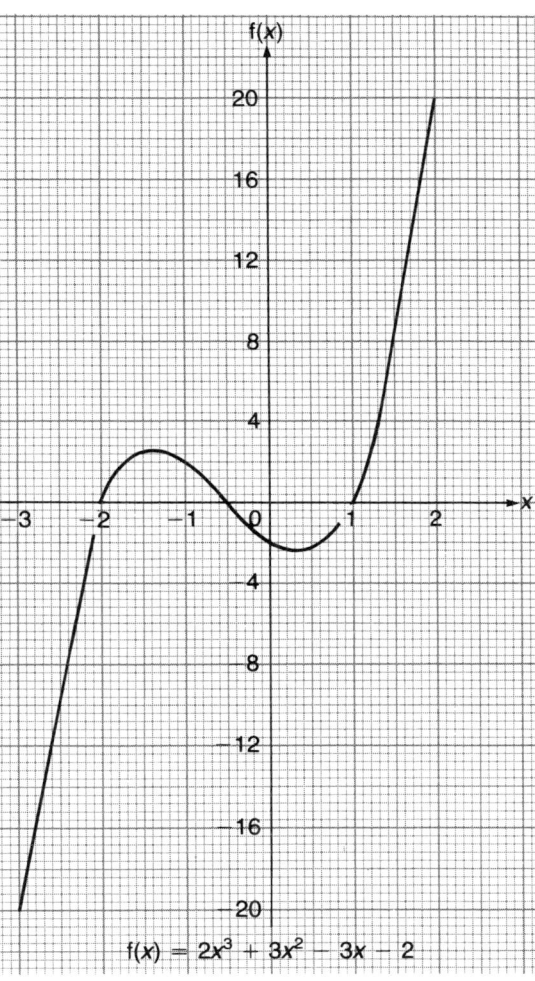

(c) Roots at $x = -2$, $x = 1$ and $x \approx -0.4$

I4 (a)

x	−5	−4	−3	−2	−1	0	1	2	3
$-x^3$	125	64	27	8	1	0	−1	−8	−27
x^2	25	16	9	4	1	0	1	4	9
$14x$	−70	−56	−42	−28	−14	0	14	28	42
-24	−24	−24	−24	−24	−24	−24	−24	−24	−24
$f(x)$	56	0	−30	−40	−36	−24	−10	0	0

(b)

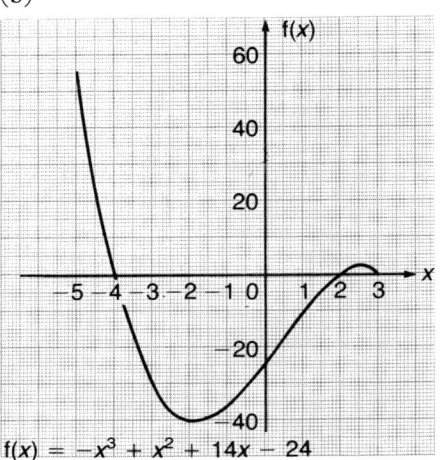

$f(x) = -x^3 + x^2 + 14x - 24$

(c) Roots at $x = -4$, $x = 2$ and $x = 3$

I5 (a)

x	−2	−1	0	1	2	3	4	5
$-2x^3$	16	2	0	−2	−16	−54	−128	−250
$10x^2$	40	10	0	10	40	90	160	250
$-4x$	8	4	0	−4	−8	−12	−16	−20
-16	−16	−16	−16	−16	−16	−16	−16	−16
$f(x)$	48	0	−16	−12	0	8	0	−36

(b)

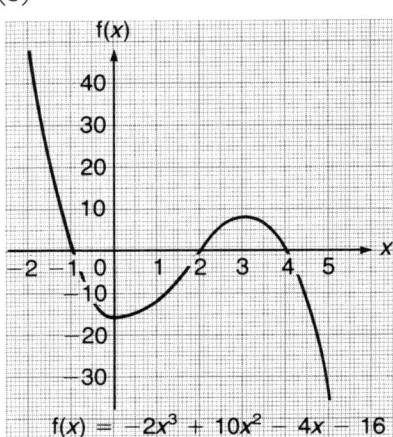

$f(x) = -2x^3 + 10x^2 - 4x - 16$

(c) Roots at $x = -1$, $x = 2$ and $x = 4$

J Making the best of it

J1 (a) Length = $(30 - 2h)$ cm
Breadth = $(20 - 2h)$ cm

(b) $V(h) = 4h^3 - 100h^2 + 600h$

(c)

h	0	2	4	6	8	10
$4h^3$	0	32	256	864	2048	4 000
$-100h^2$	0	−400	−1600	−3600	−6400	−10 000
$600h$	0	1200	2400	3600	4800	6 000
$V(h)$	0	832	1056	864	448	0

(d)

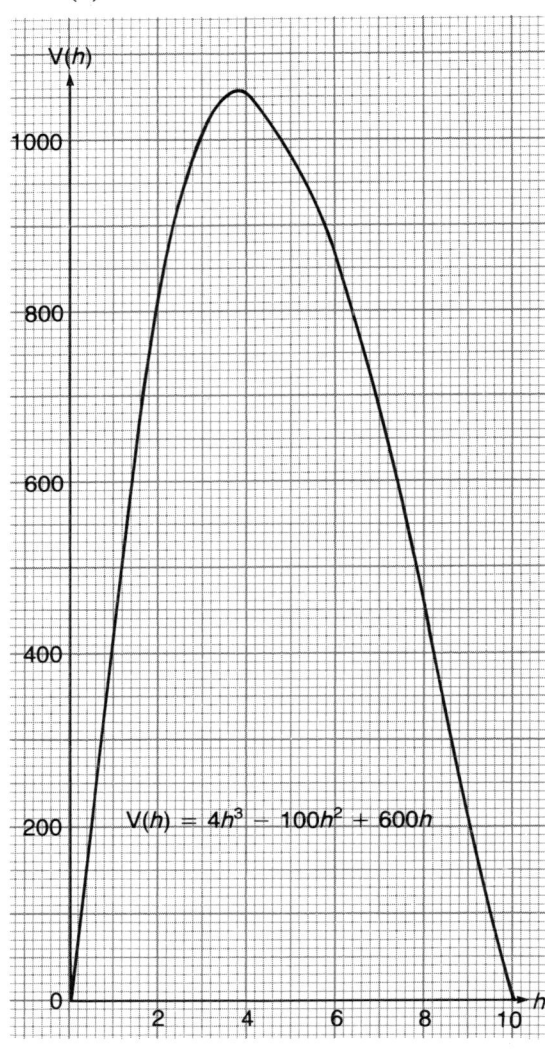

(e) Length 22 cm, breadth 12 cm, height 4 cm, volume 1056 cm³

J2 (a) Length = $(25 - x)$ cm
     Breadth = $(20 - 2x)$ cm
  (b) $V(x) = 500x - 70x^2 + 2x^3$
  (c)

x	1	2	3	4
$500x$	500	1000	1500	2000
$-70x^2$	-70	-280	-630	-1120
$2x^3$	2	16	54	128
$V(x)$	432	736	924	1008

x	5	6	7	8
$500x$	2500	3000	3500	4000
$-70x^2$	-1750	-2520	-3430	-4480
$2x^3$	250	432	686	1024
$V(x)$	1000	912	756	544

J2 (d)

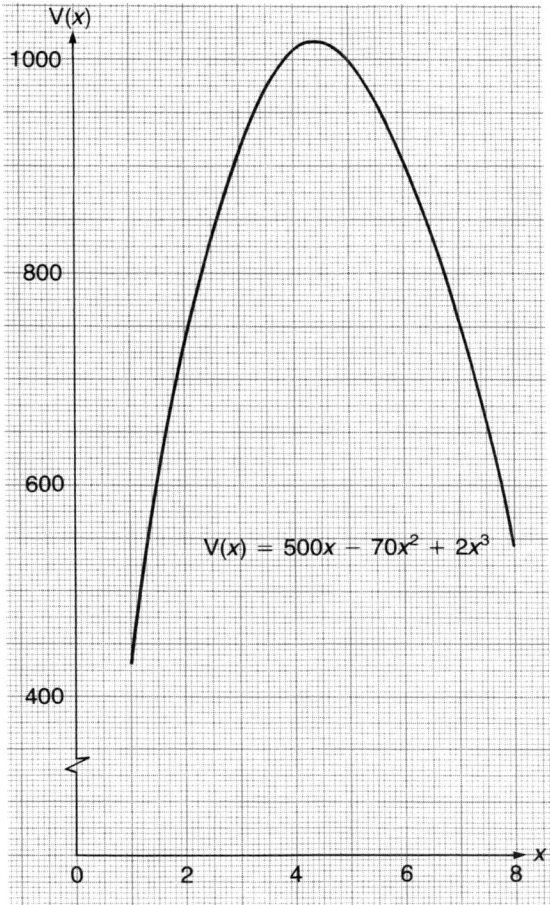

(e) Length 20·5 cm, breadth 11 cm, height 4·5 cm, volume 1014·75 cm³

Money management 3: hire purchase and VAT

This section has Cathy and Colin buying a stereo system. They look at different ways of buying it and try to decide which is best. Further examples involve calculating deposits and payments, finding pre-VAT prices, and so on.

A Hire purchase

A1 Forthdale's £230, House of Laser £224, George's Hifi £225, Fast Fred's £301·60, Largewood's mail order £239·40

A2 (a) £296 (b) £56·05 (c) 23%

A3 (a) £243·75 (b) 11%

A4 (a) £6·31 (b) £269·57

A5 £52·51

A6 Scheme C

A7 £125.82

B VAT

B1 (a) £34·50 (b) £6·33 (c) £376·28
     (d) £19·90 (e) £17

B2 (a) £260 (b) £2·61 (c) £95·65
     (d) £173·91 (e) £200·79 (f) £100·52

---

5 Angles greater than 90°

This chapter begins by using rotating-arm diagrams to find the signs of the trigonometric ratios and the related acute angles. The 'CAST' rule is not specifically mentioned in the text as pupils should formalise it for themselves. The chapter goes on to introduce simple trigonometric equations in the form of puzzles and formalises the algebra of identities.

A The trundle wheel

A1

(a) 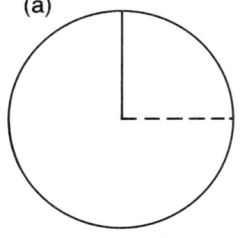 (b)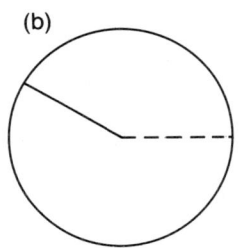

(c) 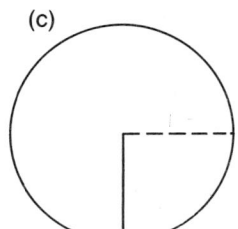 (d)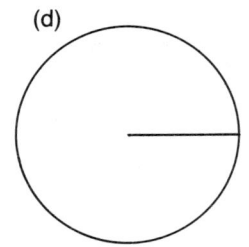

A2 (a) 60° (b) 8·7 cm

A3

Degrees from start	Above or below?	Distance MP
0	—	0 cm
30	Above	5 cm
60	Above	8·7 cm
90	Above	10 cm
120	Above	8·7 cm
150	Above	5 cm
180	—	0 cm
210	Below	5 cm
240	Below	8·7 cm
270	Below	10 cm
300	Below	8·7 cm
330	Below	5 cm
360	—	0 cm

A4

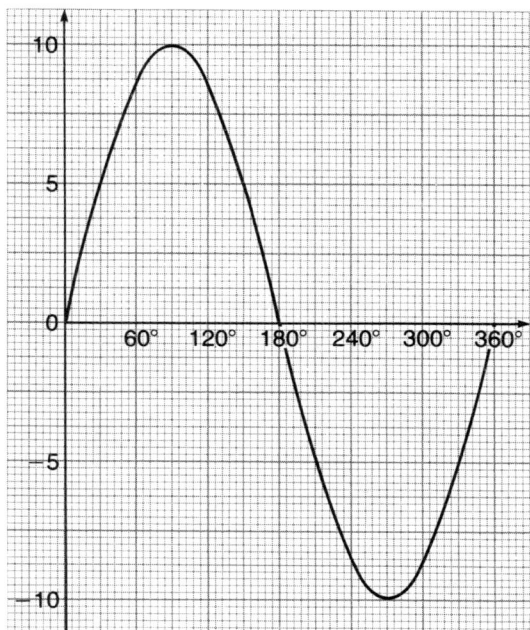

A5 (a) 9 cm (b) 1·7 cm (c) 6 cm

A6 30°, 150°, 210°, 330°

A7 About 50°, 130°, 230°, 310°

A8 148°, 212°, 328°

A9

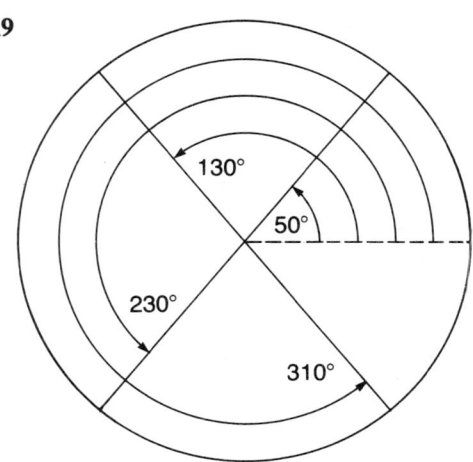

A10 Start with an acute angle $x°$, then the angles are $(180 - x)°$, $(180 + x)°$ and $(360 - x)°$.

A11 Graph would repeat itself.

A12

(a) 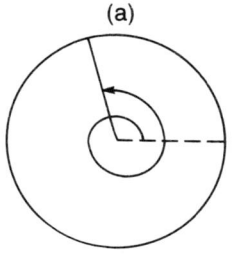   (b)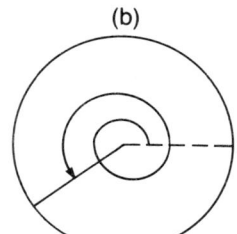

(c) 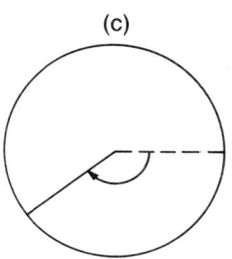   (d)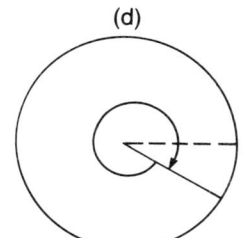

B Trigonometric ratios

B1 (a) $\sin x° = \dfrac{a}{c}$, $\cos x° = \dfrac{b}{c}$, $\tan x° = \dfrac{a}{b}$
  (b) All are positive.

B2 (a) $\sin x° = \dfrac{a}{c}$, $\cos x° = \dfrac{b}{c}$, $\tan x° = \dfrac{-a}{b}$
  (b) Sine is positive.

B3 (a) $\sin x° = \dfrac{-a}{c}$, $\cos x° = \dfrac{-b}{c}$, $\tan x° = \dfrac{a}{b}$
  (b) Tangent is positive.

B4 (a) $\sin x° = \dfrac{-a}{c}$, $\cos x° = \dfrac{b}{c}$, $\tan x° = \dfrac{-a}{b}$
  (b) Cosine is positive.

25

B5

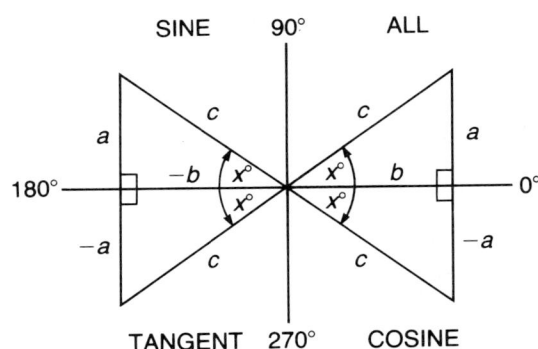

C Angles greater than 90°

C2 (a) Second quadrant
(b)
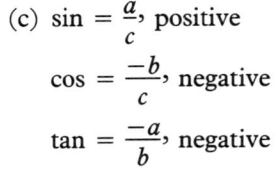

(c) $\sin = \dfrac{a}{c}$, positive

$\cos = \dfrac{-b}{c}$, negative

$\tan = \dfrac{-a}{b}$, negative

C3 (a) Fourth quadrant
(b)
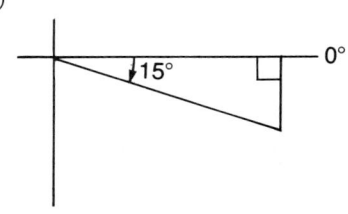

(c) $\sin = \dfrac{-a}{b}$, negative

$\cos = \dfrac{b}{c}$, positive

$\tan = \dfrac{-a}{b}$, negative

C4 (a) First quadrant
(b)
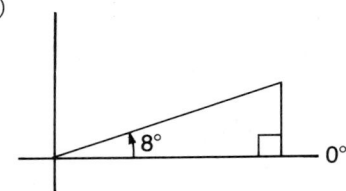

(c) $\sin = \dfrac{a}{c}$, positive

$\cos = \dfrac{b}{c}$, positive

$\tan = \dfrac{a}{c}$, positive

C5 (a) Second quadrant
(b)
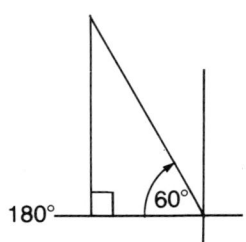

(c) $\sin = \dfrac{a}{c}$, positive

$\cos = \dfrac{-b}{c}$, negative

$\tan = \dfrac{-a}{c}$, negative

C6 (a) Third quadrant
(b)

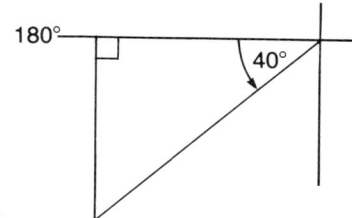

(c) $\sin = \dfrac{-a}{c}$, negative

$\cos = \dfrac{-b}{c}$, negative

$\tan = \dfrac{a}{b}$, positive

D Trigonometric ratios of angles greater than 90°

D1 (a) −0·532 (b) 0·682 (c) 0·530
(d) −0·391 (e) 11·430 (f) −0·643
(g) 3·271 (h) 0·122 (i) 0·988

E Special angles greater than 90°

E1 (a) $\dfrac{-1}{2}$ (b) $\dfrac{-1}{\sqrt{2}}$ (c) $\dfrac{1}{\sqrt{2}}$ (d) $\dfrac{1}{\sqrt{2}}$
(e) −1 (f) $\dfrac{-\sqrt{3}}{2}$

F Review

F1 (a) 20° (b) 180° − 20° = 160°
(c) Use 'bow tie' (d) 2

F2 (a) 45° (b) 180° − 45° = 135°
(c) Use 'bow tie' (d) 2

F3 (a) 2 (b) First or third
(c) 70°, 290°, 430°, 650° 1510°
(Let your teacher see your explanations.)

F4 (a) 40°, 140° (b) 95°, 275°
(c) 152°, 208° (d) 45°, 315°
(e) 120°, 300°

G Trigonometric equations

G1 (a) 47·4°, 132·6° (b) 107·4°, 287·4°
(c) 122·6°, 237·4° (d) 191·6°, 348·4°
(e) No values (f) 108·4°, 289·4°

G2 (a) 51·3°, 231·3° (b) 233·1°, 306·9°
(c) 0°, 360° (d) 53·8°, 233·8°
(e) No values

H Identities

I The algebra of identities

Show your answers to your teacher.

Money management 4: savings and investment

This section includes examples on simple interest and compound interest and on comparisons between the two. An investigation of the best savings schemes to fit Cathy and Colin's needs is also included, which may raise interesting points for class discussion.

A **Interest**

A1 (a) £342 (b) £368·75 (c) £415·30
   (d) £175 + $\left(\dfrac{1225 \times 0·0x}{12}\right)$ (e) £57·85
   (f) £457·37 (g) £366·58

A2 8%   A3 3%   A4 £10 000

A5 6 years 3 months

B **Investigation**

Show your answer to your teacher.

C **Stocks and shares**

C1 £162·50   C2 £146·25

C3 Tesbury's £10·98, Overwoods £21·11, Cowntree £24·44

6 Similar triangles

A method of finding the height of a tree, taken from the book *Scouting for Boys*, introduces this chapter and forms the basis for a consideration of similar triangles. Then, practical work on constructing and measuring triangles forms the basis for an investigation of the relationship between pairs of equiangular triangles and the ratio of sides. Having established this relationship, a more formal, traditional approach is taken to setting out a proof that triangles are similar. Pupils are asked to produce proof, to ensure that they know which angles are equal and are not just pairing them up arbitrarily. The final section deals with practical examples in context. Ratios of areas and volumes of similar shapes and solids are included in *Mathematics for Credit Book 2*.

A **Equiangular triangles**

A1

Triangle ABC	Triangle DEF
AB = 2 cm	DE = 6 cm
BC = 1·7 cm	EF = 5 cm
AC = 1·4 cm	DF = 4·2 cm

A2 × 3

A3 Show your answer to your teacher.

A4

Triangle ABC	Triangle DEF
$\hat{A}$ = 55°	$\hat{D}$ = 55°
$\hat{B}$ = 44°	$\hat{E}$ = 44°
$\hat{C}$ = 81°	$\hat{F}$ = 81°

A5 The angles are equal in pairs.

A6, A7 Show your answers to your teacher.

A8 (a) $\hat{D}$ = 60°, $\hat{E}$ = 40°
   (b) $\hat{B}$ = 110°, $\hat{C}$ = 40°, $\hat{D}$ = 30°, $\hat{F}$ = 40°
   (c) $\hat{R}$ = 20°, $\hat{Q}$ = 70°, $\hat{Y}$ = 90°, $\hat{Z}$ = 70°
   (d) $\hat{K}$ = 3x°, $\hat{M}$ = 120°, $\hat{S}$ = x°
   (e) $\hat{G}$ = 3x°, $\hat{R}$ = 40° − x°, $\hat{Q}$ = 2x°
   (f) $\hat{Z}$ = x°, $\hat{D}$ = x°, $\hat{C}$ = 110° − x°

B **Corresponding sides**

B1

Triangle ABC	Triangle DEF	Ratio
AB = 4 cm	DE = 7·5 cm	$\dfrac{AB}{DE}$ = 0·53
BC = 3·1 cm	EF = 5·8 cm	$\dfrac{BC}{EF}$ = 0·53
AC = 2·8 cm	FD = 5·3 cm	$\dfrac{AC}{FD}$ = 0·53

B2 In two equiangular triangles, the ratios of the sides opposite the equal angles **are equal**.

B3, B4, B5 Show your answers to your teacher.

B6 (a)

Corresponding sides	Opposite angle
PA and QT	50°
PB and QR	60°
AB and TR	70°

(b)

Corresponding sides	Opposite angle
PT and US	40°
KP and LU	25°
KT and LS	115°

(c)

Corresponding sides	Opposite angle
XY and GF	45°
PY and MF	65°
XP and GM	70°

(d)

Corresponding sides	Opposite angle
NS and ZR	●
SW and RQ	○
NW and ZQ	⌒

(e)

Corresponding sides	Opposite angle
BD and XH	×
BG and XL	●
DG and HL	⌒

C Using similar triangles to show that angles are equal

C1 Show your answer to your teacher.

C2 (a) Not similar (b) Not similar
(c) $\hat{Z} = \hat{E}, \hat{X} = \hat{D}, \hat{Y} = \hat{F}$
(d) Not similar

D Using similar triangles to calculate sides

D1 RP = 12·03 cm **D2** YZ = 15 cm

D3 QR = 4·2 cm

D4 (a) AE = 30 m (b) RS = 6 cm
(c) KL = 5 m (d) PR = 6 cm
(e) EF = 15 cm (f) VY = 5·625 cm

E Using similar triangles to solve real problems

E1 1·05 m **E2** 63 cm **E3** 2 m

E4 35 cm **E5** 13·33 m

Consolidation 2

Graphs (1)

1 (a)

x	−1	0	1	2	3
x^2	1	0	1	4	9
$-2x$	2	0	−2	−4	−6
4	4	4	4	4	4
$f(x)$	7	4	3	4	7

(b)

x	−1	0	1	2	3
x^2	1	0	1	4	9
$-3x$	3	0	−3	−6	−9
-10	−10	−10	−10	−10	−10
$f(x)$	−6	−10	−12	−12	−10

(c)

x	−1	0	1	2	3
$3x^2$	3	0	3	12	27
$2x$	−2	0	2	4	6
-7	−7	−7	−7	−7	−7
$f(x)$	−6	−7	−2	9	26

(d)

x	−1	0	1	2	3
$-4x^2$	−4	0	−4	−16	−36
$-9x$	9	0	−9	−18	−27
5	5	5	5	5	5
$f(x)$	10	5	−8	−29	−58

2 (a)

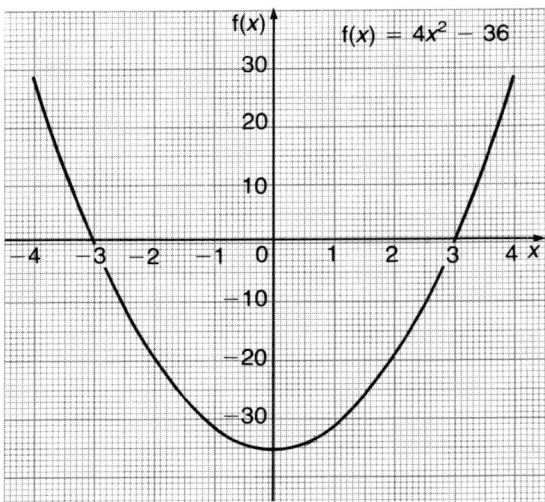

(b)

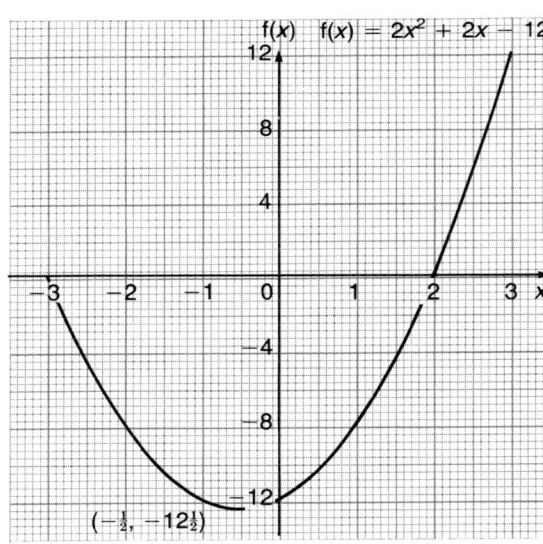

3 At $x = -5$ and $x = 4$ 4 49

5 $-12\frac{1}{4}$

6 (a) $A(x) = 4x^2 + 20x$
   (b)

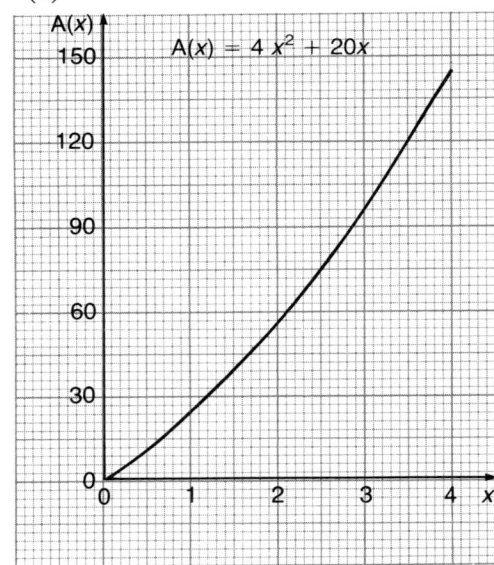

(c) $x = 1$

Hire purchase and VAT

1 £70·01   2 £220·24

3 (a) £10·56   (b) £554   (c) Scheme (b)

4 (a) VAT £1·19, Total £9·14
  (b) VAT £1·52, Total £11·67

5 (a) DIY materials £16·54, Total £19·02
  (b) Materials £7·26, VAT £1·08

Angles greater than 90°

1 (a) 39·2°, 140·8°   (b) 103·4°, 256·6°
  (c) 101·5°, 281·5°   (d) 10·6°, 349·4°
  (e) 45°, 135°   (f) 150°, 210°
  (g) 120°, 300°   (h) 30°, 150°

3 (a) $x = 75·5°$ or $284·5°$
   (b) $x = 35°$ or $145°$
   (c) $x = 108·2°$ or $251·8°$
   (d) $x = 116·6°$ or $296·6°$
   (e) $x = 181·9°$ or $358·1°$
   (f) $x = 109·5°$ or $250·5°$
   (g) $x = 35·3°$ or $215·3°$
   (h) $x = 120°$ or $240°$
   (i) $x = 180°$
   (j) $x = 60°$ or $120°$

Savings and investment

1 Account C

2 (a) 228 (b) £27·79

Similar triangles

1 (a) AB = 10 cm
   (b) PR = 5·25 cm, QR = 6 cm

2 Small triangle: side 5·42 cm, angles 42° and 73°
   Large triangle: side 8·58 cm, angles 65° and 73°

3 (a) 4 : 5 (b) 16 : 25 (c) 64 : 125

7 The cosine rule

We have introduced this rule before the sine rule so that pupils will have a clear understanding of its mechanics, rather than using it only when the sine rule fails.

Working through real examples paves the way for the proof of the cosine rule, which is felt should be included for completeness. Once the formula has been established, it is emphasised using words, and some oral work may be appropriate here. Similarly, words are again used to highlight the version of the rule where the cosine is the subject. There is no specific mention of the 'two sides and the included angle' rule in the text, as the pupils should discover this themselves in an informal way; teachers may wish to highlight this point further.

Clearly, if results are to be given to 3 s.f., all calculations should be to at least 4 s.f., with rounding to 3 s.f. at the end of the problem.

A The problem

A1 $\sin 40° = \dfrac{CD}{AC}$ $\cos 40° = \dfrac{AD}{AC}$

$0·643 = \dfrac{CD}{12}$ $0·766 = \dfrac{AD}{12}$

CD = 7·72 cm   AD = 9·19 cm

A2 5·81 cm   **A3** 9·66 cm

B The obtuse-angled case

B1 55°

B2 $\sin 55° = \dfrac{CD}{AC}$ $\cos 55° = \dfrac{AD}{AC}$

$0·819 = \dfrac{CD}{12}$ $0·574 = \dfrac{AD}{12}$

CD = 9·83 cm   AD = 6·89 cm

B3 21·8 cm   **B4** 23·9 cm

C The right-angled case

C1 19·2 cm

C2 $a^2 = b^2 + c^2 - 2bc \cos \hat{A}$
 $= 12^2 + 15^2 - (2 \times 12 \times 15 \times \cos 90°)$
 $= 144 + 225 - (2 \times 12 \times 15 \times 0)$
 $= 369$
 $a = 19·2$ cm
 Value of a obtained in question C1 = 19·2 cm.
 Formula holds.

D Using the formula to find the side of a triangle

D1 (a) $b^2 = a^2 + c^2 - 2ac \cos \hat{B}$
 (b) $c^2 = a^2 + b^2 - 2ab \cos \hat{C}$
 (c) $p^2 = q^2 + r^2 - 2qr \cos \hat{P}$
 (d) $e^2 = d^2 + f^2 - 2df \cos \hat{E}$
 (e) $z^2 = x^2 + y^2 - 2xy \cos \hat{Z}$
 (f) $f^2 = a^2 + t^2 - 2at \cos \hat{F}$
 (g) $s^2 = y^2 + m^2 - 2ym \cos \hat{S}$

D2 (a) 9 cm (b) 4·45 m (c) 8.35 cm
 (d) 30.5 cm (e) 12·6 cm
 (f) 6·39 cm

D3 (a) 3·75 cm (b) 5·36 m (c) 3·75 m

D4 37·65 cm   **D5** 181 km

D6 200 km   **D7** 10·5 km

E Finding an angle in a triangle, given its three sides

E1 (a) $\cos \hat{B} = \dfrac{a^2 + c^2 - b^2}{2ac}$
 (b) $\cos \hat{C} = \dfrac{a^2 + b^2 - c^2}{2ab}$
 (c) $\cos \hat{P} = \dfrac{t^2 + r^2 - p^2}{2tr}$
 (d) $\cos \hat{D} = \dfrac{a^2 + t^2 - d^2}{2at}$
 (e) $\cos \hat{Q} = \dfrac{b^2 + a^2 - q^2}{2ab}$
 (f) $\cos \hat{H} = \dfrac{g^2 + b^2 - h^2}{2bg}$

E2 (a) 59·5° (b) 29·9° (c) 72° (d) 68·5°
 (e) 93·9° (f) 145·4°

E3 10·3°   **E4** 73·9°

E5 (a) 86·6° (b) 87·2°

Money management 5: loans and insurance

These topics are introduced through Colin buying a car and insuring it; the section then goes on to look at other types of loan and insurance. Once again, much emphasis is put on comparing different schemes and choosing the best one to suit a given set of needs.

A Loans

A1 £113·97

A2 (a) £92·20 (b) £200·69

A3 (a) £188·81 (b) £4180·47

A4 Show your answer to your teacher.

B Other types of loan

B1 (a) £396·61 (b) £38·80

B2 Show your answer to your teacher.

C Car insurance

C1 (a) Compensation to the injured pedestrian
 (b) Cost of repairs to the man's car
 (c) Cost of repairs to both cars

C2 £140   **C3** £300

C4 (a) 40% (b) £184·50

C5 (a) £195 (b) £243·75
 (c) Pay for repairs, since this would be cheaper than losing his no claims discount.

D House insurance

D1 (a) £135·45 (b) £66·20 (c) £375·30

D2 House £56 800, contents £25 000

D3 £104 000

8 Patterns

This chapter is an ideal accompaniment to pattern type investigations; it looks at simple sequences and at how a general rule can be found to calculate any term of the sequence (the nth term rule). Flow charts are also used to generate the terms of a sequence — another example of an iterative process. Towards the end of the chapter is a section which leads the pupil through the finding of the nth term of the sequence of triangular numbers and another which deals with sequences within a practical context.

A Sequences

A1 (a)

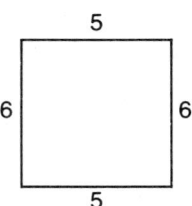

(b) 10, 12

(c)

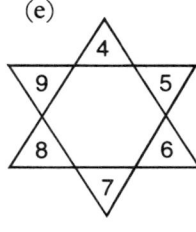 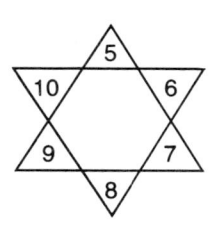

(d) 16, 21

(e)

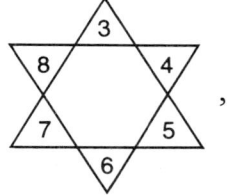

(f) 30, 31

A2 3, 6, 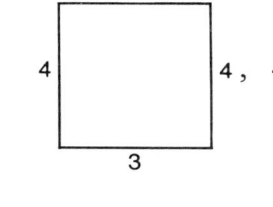, 31

A3 (a) $u_7 = 13$, Add 2
 (b) $u_6 = 2160$, Multiply by one extra
 (c) $u_4 = \frac{1}{16}$, Multiply by $\frac{1}{2}$
 (d) $u_8 = -0.6$, Subtract 0·2

A4 (a) Add 5 (b) 7 (c) $u_8 = 40$

A5 (a) Subtract 3 (b) 24 (c) $u_{12} = 67$

A6 £140   **A7** £1822·50

B Flow charts

B1 $u_1 = 1, u_2 = 3, u_3 = 5, u_4 = 7, u_5 = 9$

B2 $u_1 = 5, u_2 = 4·5, u_3 = 4, u_4 = 3·5, u_5 = 3,$
 $u_6 = 2·5, u_7 = 2, u_8 = 1·5$

B3 $u_1 = 2, u_2 = 9, u_3 = 44, u_4 = 219,$
 $u_5 = 1095, u_6 = 5469, u_7 = 27\,344$

B4 $u_1 = 1, u_2 = 9, u_3 = 33, u_4 = 105, u_5 = 321$

C The 'any-term' rule

C1 $u_9 = 81$, $u_{25} = 625$, $u_{63} = 3969$

C2 $u_n = n^2$

C3 (a) $u_{57} = 171$, $u_{949} = 2847$, $u_{1245} = 3735$
(b) $u_n = 3n$

C4 (a) $u_n = 6n$ (b) $u_{24} = 144$ (c) u_{15}

C5 (a) $u_n = n + 4$ (b) $u_{156} = 160$

C6 2, 5, 8, 11, 14

C7 2, 6, 12, 20

C8 $u_5 = 32$, $u_{12} = 4096$

C9 $u_1 = 1$, $u_3 = 7$, $u_5 = 21$, $u_8 = 57$

C10 (a) $u_n = 4n - 1$ (b) $u_n = n^3$
(c) $u_n = 3n + 2$ (d) $u_n = \frac{1}{2}n$
(e) $u_n = n^2 + n$ or $n(n + 1)$

D Problems

D1 $u_n = 8n - 1$

D2 (a) 1, 9, 25, 49 (b) $u_5 = 81$, $u_6 = 121$
(c) Squares of the sequence of odd numbers
(d) $u_n = (2n - 1)^2$

E Triangular numbers

E1 $u_{15} = 120$, $u_{20} = 210$

E2

Length	11	38	$n + 1$
Breadth	10	37	n

E3

Position	3	4	10	37	n
Dots	12	20	110	1046	$n^2 + n$

E4 Triangular numbers are half the number of dots.

E5 $u_n = \frac{1}{2}n(n + 1)$

E6 $u_{75} = 2850$, $u_{130} = 8515$

E7 (a) 60th (b) 31st

F Investigations

F1 (a) 20 matches
(b) Show your answer to your teacher.
(c) $n(n - 1)$ matches, if n is the number of players
(d) Show your answer to your teacher.

F2 (a) 6
(b), (c) Show your answers to your teacher.

9 Variation

This chapter starts where the work of SMP *Book Y3* leaves off, translating the terms of 'proportionality' into the language of variation. Examples are given of both direct and inverse variation involving evaluation and sketching graphs. Variation involving several variables is also recognised and is dealt with by searching for a straight line relationship (for example, between y and $\frac{1}{x}$). Much importance is placed on determining the consequences of changing one of the variables. Examples are set in context wherever possible.

A Direct variation

A1 (a) $d \propto t$ (b) $k = 105$ (c) $d = 105t$
 (d) 78·75 km (e) Halved
 (f)

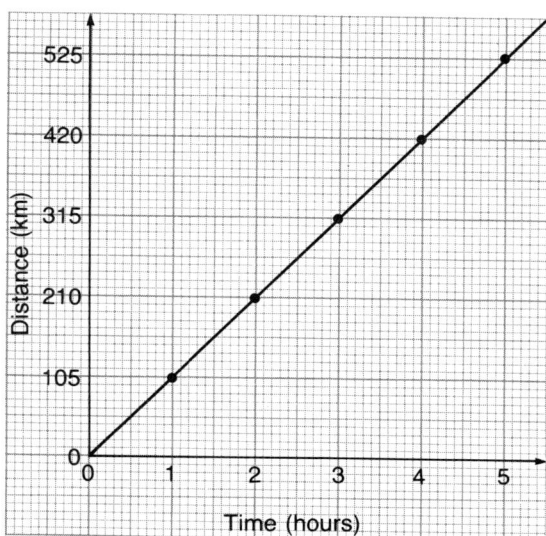

A2 (a) $C \propto r$ (b) 13π cm (c) 1·75 cm
 (d) Doubled (e) Halved
 (f)

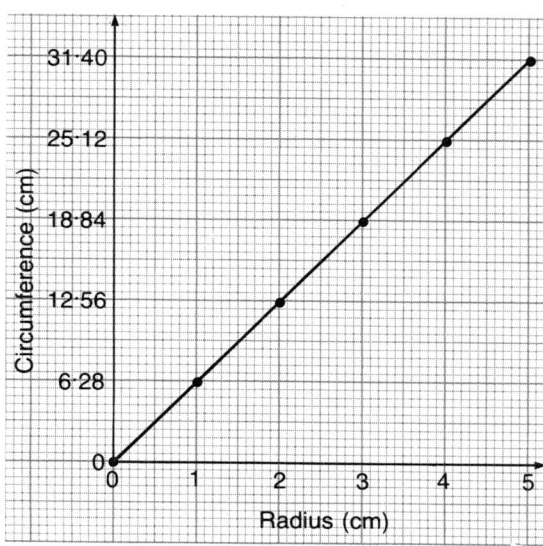

A3 (a) $k = 5$ (b) $P = 5l$
 (c) Reduced by 25%
 (d) Reduced by 20%

A3 (e)

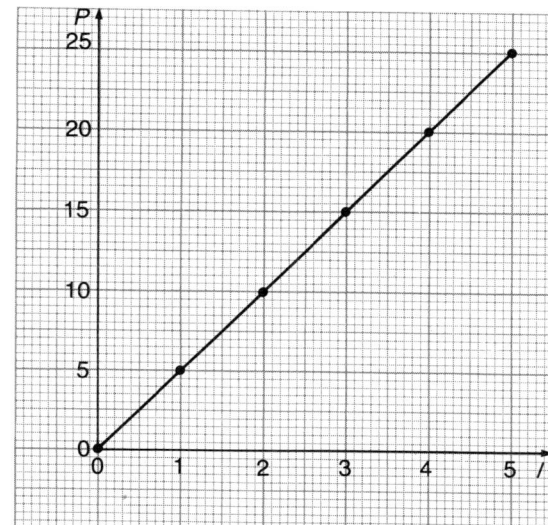

A4 (a) $k = \frac{1}{6}$, $I = \frac{1}{6}V$
 (b) (i) $I = 0·05$ A (ii) $V = 75$ V

B More interesting direct variation

B1 (a)

Length of side	1	2	3	4	...	10
Area of top	1	4	9	16	...	100
Weight	20	80	180	320	...	2000

 (b) $W \propto l^2$, $k = 20$ (c) $W = 20l^2$
 (d) 2880 g (e) 4 cm
 (f) Multiplied by 4
 (g)

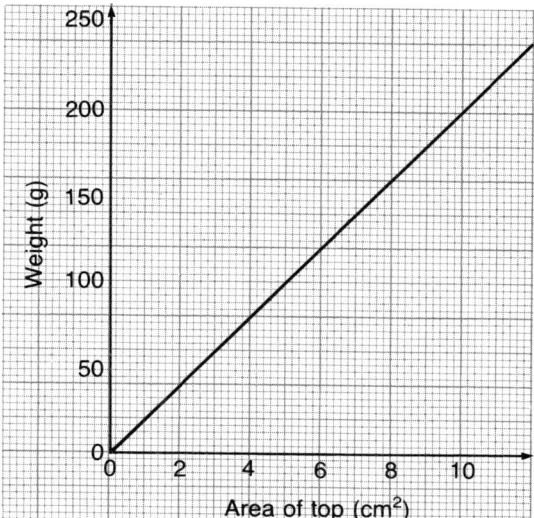

B2 (a) 200 cm² (b) 8 times bigger

B3 (a) 15·625 times bigger
(b) 4 times bigger
(c) Volume is $\frac{27}{64}$ of its original value.

B4 (a) $k = 0·2$ (b) 25 cm
(c) Lengthened by 36 cm

C Inverse variation

C1 (a) $k = 77·48$, $X = \dfrac{77·48}{Y}$ (b) $Y = 9·685$
(c)

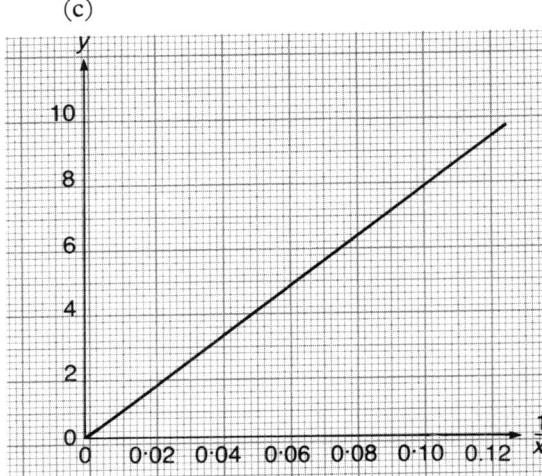

C2 (a) $k = 3$

a	2	3	4	5	6
b	1·5	1	0·75	0·6	0·5

(b) Divided by 3 (c) Doubled
(d)

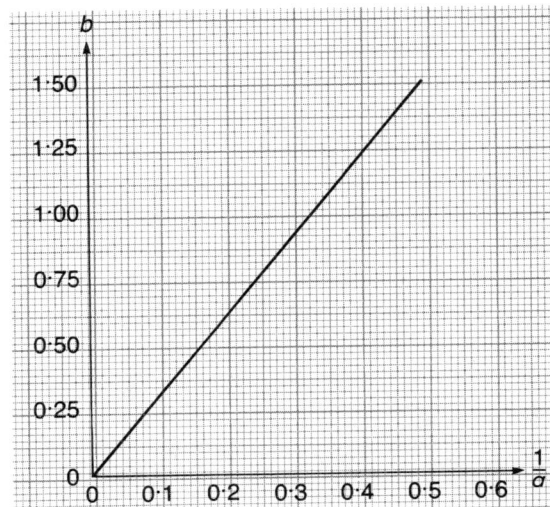

C3 (a) $k = 300\,000$
(b)

Radio station	Wavelength	Frequency
Radio Clyde	261	1150
Radio 1	285	1153
Radio Luxembourg	208	1442

(c) 909 kHz, by 103 m (d) 4% higher

C4 (a) 7 people (b) 53 people (c) £3 each
(d)

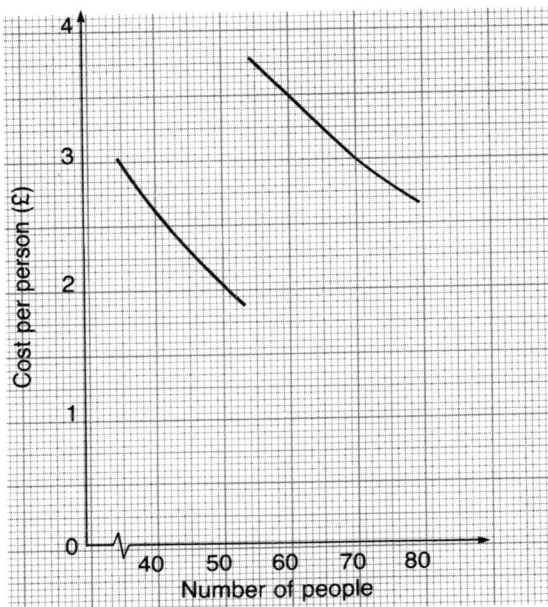

(e) 54 people

D More interesting inverse variation

D1 (a)

l	3	6	8	12
l^2	9	36	64	144
h	40	10	5·625	2·5
v	120	120	120	120

D1 (b)

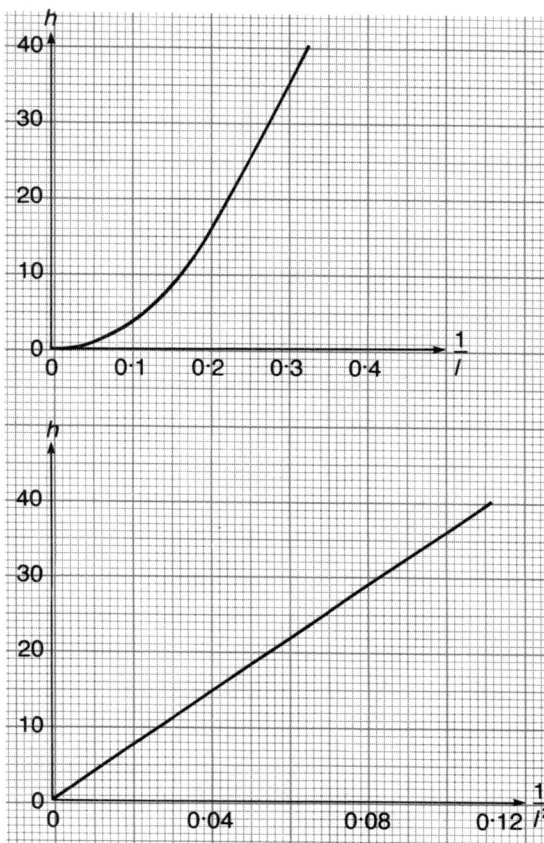

(c) $h \propto \dfrac{1}{l^2}$ (d) $k = 120$

(e) 22·5 cm (f) Multiplied by 4

D2 (a) 200π cm³

(b) Graph of $\left(\dfrac{1}{r^2}, h\right)$ is a straight line.

(c) $k = 600$ (d) Multiplied by $\tfrac{1}{16}$

D3 (a) $y = 3$ (b) $x = 0.25$

D4 (a) $A = \dfrac{3\cdot 2}{b^3}$ (b) Reduced by $\tfrac{7}{8}$

D5 (a) Multiplied by 4 (b) Reduced by $\tfrac{3}{4}$
(c) Multiplied by $\tfrac{16}{9}$

D6 (a) $\tfrac{4}{9}$ (b) 6400 km

E A variation on variation

E1 (a) Direct, $k = 5$ (b) Inverse, $k = 18$
(c) Inverse, $k = 6\cdot 4$ (d) Direct, $k = 500$

E2 (a) $y \propto \dfrac{1}{x^2}$, $k = 32$ (b) $y \propto \sqrt{x}$, $k = 16$
(c) $y \propto x$, $k = 10$ (d) $y \propto x^2$, $k = 0\cdot 5$
(e) $y \propto \dfrac{1}{x^2}$, $k = 3\cdot 6$

F Joint variation

F1 (a)

bh	6	6	6
l	1	2	3
V	6	12	18

(b)

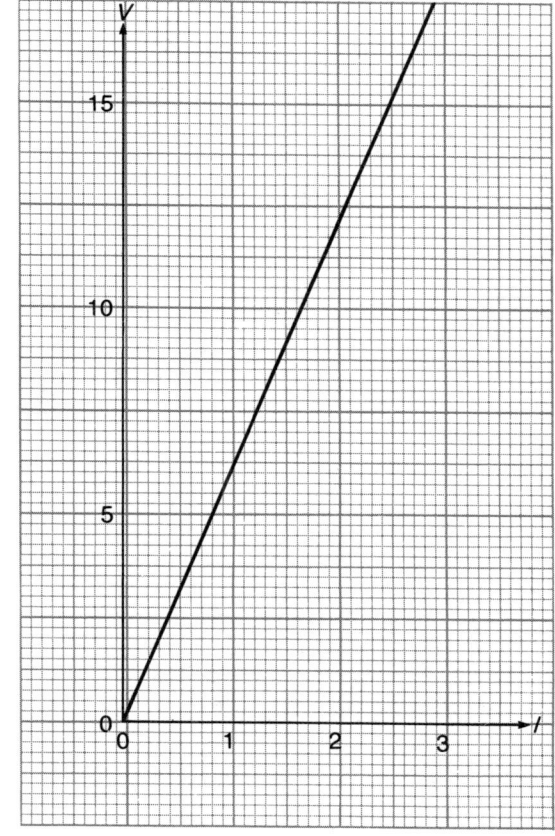

(c) $V \propto l$

F2 (a)

lh	3	3	3
b	1	4	6
V	3	12	18

F2 (b)

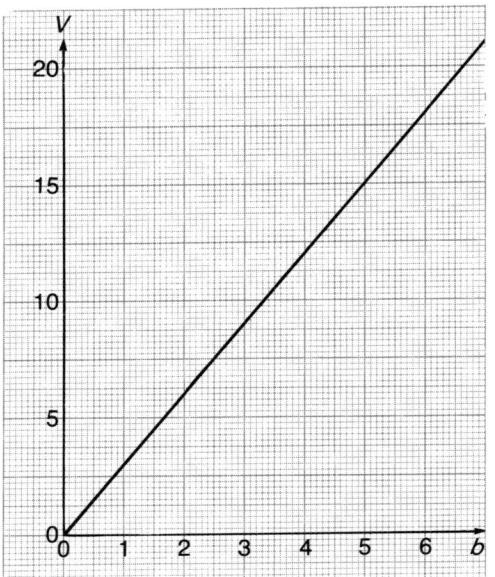

(c) $V \propto b$

F3 (a)

l	1	2	3
b	2	4	6
h	3	3	3
lb	2	8	18
V	6	24	48

(b)

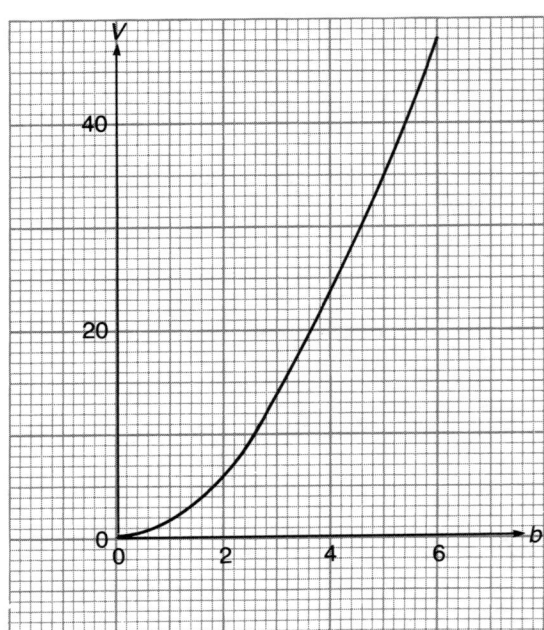

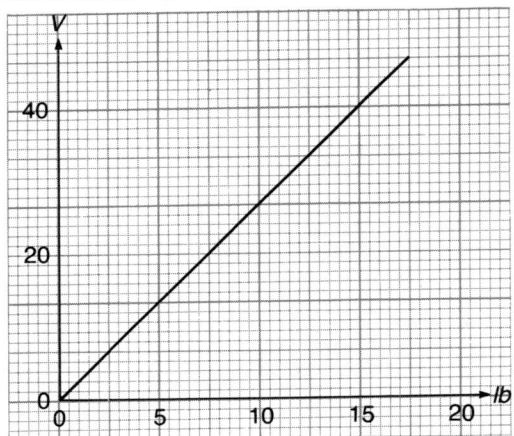

(c) $V \propto lb$, $k = 3$

F4 (a) $k = 2$ (b) $V = 2lb$ (c) $V = 48$ cm^3
  (d) Doubled

F5 (a) $k = 3, X = 3YZ$ (b) $Z = 1$
  (c) Multiplied by 1·5

F6 (a) $k = 1·5, A = \dfrac{1·5B}{C}$ (b) Multiplied by 4

F7 (a) $k = \pi$
  (b) (i) Doubled (ii) Quartered

F8 (a) Reduced by $\frac{1}{3}$ (b) 50%

F9 (a) Doubled (b) 56·25%

G A variety of variation

G1 (a) Increased by 100%
(b) Reduced by 20%

G2 Multiplied by $\frac{7}{2}$

G3 1.52×10^8 km^2

G4 263%

G5 59%

G6 200 m

Consolidation 3

The cosine rule

1. $106.6°, 48.2°, 25.2°$
2. $AB = 14$ cm
3. $XZ = 8$ cm, $\hat{Z} = 72°$, $\hat{X} = 36°$
4. 620 m
5. $92.9°$

Loans and insurance

1. £155·52
2. (a) £108 (b) £111·69
3. £94·19
4. Houses in the city are more likely to be broken into.
5. £34·97
6. £131·60

Patterns

1. (a) Add 2 (b) Subtract 5 (c) Add 0·4 (d) Multiply by 3
2. (a) $u_n = 2n$ (b) $u_n = \dfrac{n}{10}$ (c) $u_n = 3^n$ (d) $u_n = 5n - 1$
3. (a) $u_5 = -8, u_9 = -20, u_{32} = -89$ (b) u_{10}
4. $u_1 = 4, u_2 = 10, u_3 = 20, u_4 = 34$

Variation

1. $x = 422$
2. (a) $k = 3.36$ (b) 0.96 m^3
3. Stays the same
4. (a) $a = \dfrac{20b}{c}$ (b) Multiplied by $\frac{5}{2}$
5. 225 minutes
6. 224·5 days